MOTOMI KAWAKAMI

Design with Precision and Flexibility

amus arts press

amus arts press
B1 Ohe Building 2-8-1 Nishitenma
Kita-ku Osaka-shi 530-0047 JAPAN
phone : +81 6 4709 7082 fax : +81 6 4709 7084

First published 2001

Design by Ryohei Kojima Design Office
Ryohei Kojima + Sayuri Watanabe
Translated by Keiko Ihara+Rie Imahori

ISBN4-946483-63-2

Printed in Japan

MOTOMI KAWAKAMI

Design with Precision and Flexibility

Bridging Technology and Humanity
–New Design Roles in a Sustainable Culture

As a design student, I had a fortunate encounter with a book. This book consisted of the collected works of Angelo Mangiarotti, the architect whom I later came to look up to as my teacher, and in whose studio I would one day work to gain valuable experience. The works in the book were diverse, but had a superb unity and unyielding identity and spirit running through all of them that moved me deeply. Their aesthetic origins in Greek classics, the humanism with a background rooted in his vision of human history as a tapestry of different peoples; it was all so overwhelming. This must have been the event in my life that gave me the confidence to unhesitatingly pursue a career in design.

I do not think that design is something which rootlessly floats along with the changing times. For me, it begins with one person and, while transcending the individual to reach a universality, nonetheless maintains its original individual identity. Design is an artistic work and an industrial product in one, and must have the capacity to satisfy conflicting requirements such as style, function, and consideration for the context and situation. However, it is this very capacity which leads the relationship between people, between people and artifacts, and between people, space and time to a state of comfort. The kind of design I want to pursue does not speak too loudly or show off meaningless styles, but has sophistication, dignity and quality, and can provide serenity and evoke emotion in people.

But I must also accept the fact that design must always make statements based on the logics, sensibilities and technologies of the times, neither can it escape the economic and social circumstances. Even the works I have produced in my short career need to be re-evaluated and corrected in this new century. The best choices I could make at the time, in material or concept, are now seen under a new light and new criteria.

One source of the new criteria comes from the now apparent "limit to growth" of this earth. The need for us to construct a society which places high value on the mentality of using an object for its full life span and to recycle, is urgent. This is as large an assignment for myself as it is for anyone else, for I will continue to design material things. Up to the time of my parents' generation, the skill to use and re-use things to their full potential, and to reduce ecological burdens was passed down through the generations. But that was in the old days, and what is gone will not return. Today's global society imposes far more complicated and serious tasks on designers.

If there is one new thing that I want to incorporate into my work while not forgetting my love for the past, it is to reinforce new exchanges between different fields of specialists. This will not only produce higher performance and quality of the designed object, but also develop tangible and intangible relationships and systems, while producing successful outcomes in which public and other design based upon recycling and re-use of resources, as well as universal solutions for all, function effortlessly.

By further expanding such exchanges, I intend not to give up my dream, which is to play a role as liaison between such divergently specialist fields as traditional craft, international problems, psychopathology, public projects, economics, space development, advanced medical technology and food etc.

Because this is indeed what ultimately leads to the pursuit of the fundamental theme of design, which is how to articulate human reason and aesthetics through fixing one's eyes on the currents of the times.

Motomi Kawakami

Motomi Kawakami is without doubt one of the designers of his generation, both in Japan and internationally, who have maintained in their work a consistent level of quality and precision, without succumbing to the influences of the various ephemeral "isms". The character of his work is confirmed not only in his design but also in the diversity of his experiences, through which he has never relinquished an authenticity of language.

He and I share a common characteristic that produces a great affinity between our two ways of confronting this profession. But we must remember that a mentor is a true mentor only if, thanks to greater experience accumulated over time, he contributes to the formation of his pupil's own creative personality. I consider it important to underline this fact, because in the East the mentor has always had a different function compared to that in the West, where still today the pupil's statement is often conditioned by the mentor's language. Even in this regard, the relationship between my work and Motomi's finds its own precise, interesting position, where quality is found above all in precision and rational discipline.

Thus in our work the domain which was handed-down through tradition and the active intake of practicality coexist, resulting in the creation of an outcome combining extremely free creativity with essential functionality.

When I look back, for more than 40 years young Japanese designers have always been present in my studio. Since the years 1966 to 69 when Motomi Kawakami was in my studio in Milan, collaborating on various themes, our mutual esteem and friendship has never had a moment of weakness.

I wish Motomi continued success, and believe that the quality of his work will remain high while providing a meaningful and important model for us.

With esteem and friendship,
Angelo Mangiarotti

As a result of industry-centered marketing management which places too much weight on economic efficiency, the diverse powers of design have been narrowed and there is a sense that it has hit a dead end. Now is the time to return to the original meaning of design as it naturally evolved from people's needs, and to re-evaluate the activity of design as one of the basic aspects of human culture. With deep thought, we should reconsider design as a means of connecting everyday life with industry and the global environment, design to transcend existing borders and urge mutual intervention between different fields of professionalism and industry.

Although the Cartesian concept of mechanical control over nature is still ruling us in many respects, we are in the process of escaping this one-sided view and embracing the concept of people in a symbiosis with all living things. From this point of view, I would like to be associated with nature in a modest manner, while having a sensitivity towards design that is deep-rooted in a view of nature which transcends mere romanticism.

Wood Sampling (91 Different Kinds of Wood – from the exhibition "21st Century Design and Environment")

The limitation of natural resources has been realized for some time now, and the call for an ecological and sustainable culture is louder than ever. Unlike fossil resources which will be spent and never replaced, forests can be well maintained through the cycle of appropriate cutting and replanting. Thus, trees are the only resource which maintains its own life cycle through such proper utilization by us, while the expressions of natural wood are extremely rich and diverse.

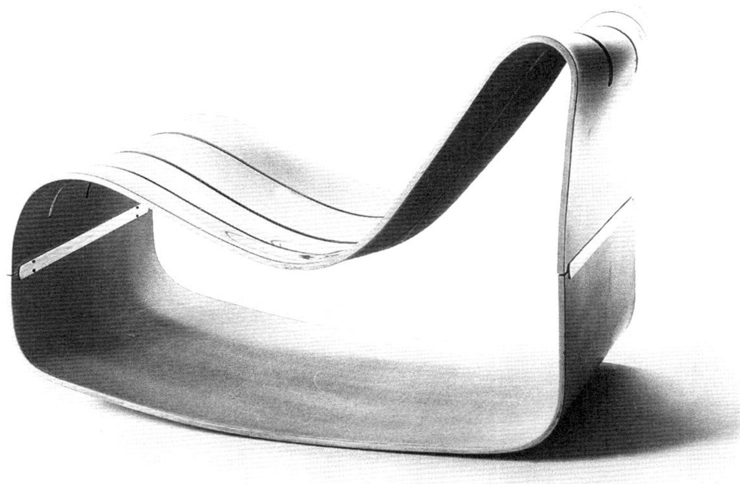

FIORENZA 1968 Alberto Bazzani s.p.a.

The subjects of my work vary from small items that fit in the palm of the hand, larger objects to support the body, to the surrounding spaces and elements of urban environments such as streets, bridges and plazas. In order to prevent existing borders between professional fields from severing my own vision, I try to move my viewpoint back and forth between the micro and macro scale of things, while always returning to the human scale as a point of reference. Among such diverse objects, the scale of furniture and especially chairs seem to fit me most naturally. Their physical scale which is close to the human body, is the meeting point of product design and environmental design, and here lies the starting point of my design work.

NT
1976
Arflex Japan Ltd.

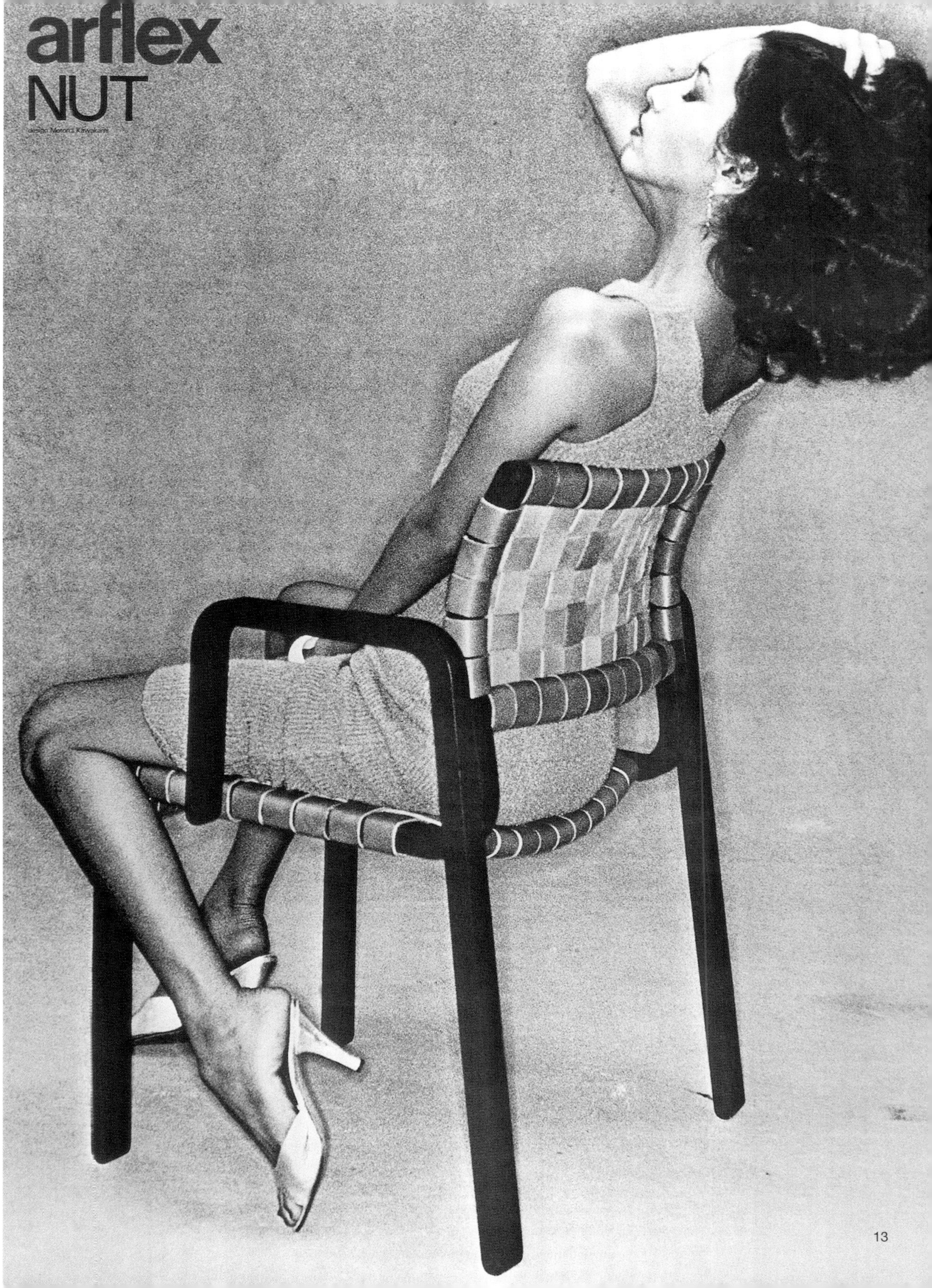
arflex
NUT
design Motomi Kawakami

MK
1992
Arflex Japan Ltd.

LUCA
1997
Arflex Japan Ltd.

ASHI (Table) MUSA (Chair)
1992 1991
Acerbis International s.p.a.

An Alternative Seating Concept: Furnishing for Public Spaces

The concept of seating is enormously diverse in the hierarchy constituted by man, object and space. For example, when we compare a working chair with a fixed seating built into the niche of a wall, we feel a great distance between them regarding their meaning rather than their common purpose, to support the human body.

In private spaces, a tactile quality, warm and intimate to the human body is required in interior design and objects. When designing chairs to be used in everyday life, I start from a tactile feeling and size on a human scale. I concentrate my senses to delve into details, then widen my vision to include the surrounding interiors, repeat this concentration and expansion of scale back and forth while making study models until I reach a state of stability.

The requirements of foyers and lobbies of office buildings or public facilities are a different story. The interior design here will have to accommodate large numbers of people and various objectives and at times serve to create a ceremonial or formal atmosphere as the public face of an enterprise. Here the design of chairs and seating will identify itself with the architectural space, sharing the same objective and context, atmosphere and meaning. Whether designing chairs to stand out and attract attention or to be merged into the surrounding interiors, their texture, color, lighting design and functional spacing, will all concentrate toward a contextual unification in this public space and its overall concept.

TICINO (Sopra Fiume Series)
1991
Okamura Corp.

OMBRONE (Sopra Fiume Series)
1991
Okamura Corp.

OFANTO (Sopra Fiume Series)
1991
Okamura Corp.

PIAVE (Sopra Fiume Series)
1991
Okamura Corp.

VOLTURNO (Sopra Fiume Series)
1991
Okamura Corp.

PLATANI (Sopra Fiume Series)
1991
Okamura Corp.

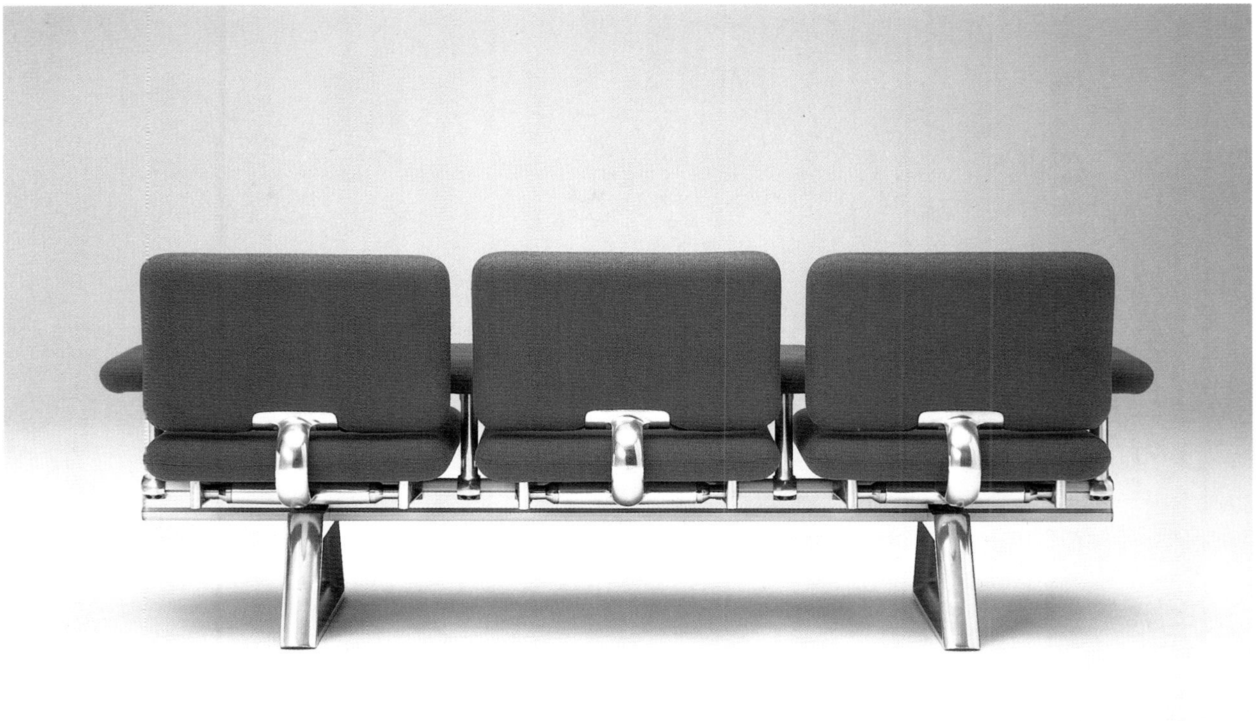

METAURO (Sopra Fiume Series)
1991
Okamura Corp.

The "GACHI" Series

While keeping in mind the practical and social realities of everyday life, a step back from the immediate requirements of efficiency and function will allow us to confront products and furniture with a fresh eye. As there are formal and casual elements in people's lives and activities, similarly there are differences between the rigid, flexible, and casual styles in furniture and artifacts also. In an attempt to represent the formal and orthodox styles of human attitude, I have tried in this piece to symbolize abstraction and spirituality, like that condensed in the Japanese "Toko-no-ma (alcove)". Here the lyricism of the changing expressions of wood surface, the hard texture of metal, the decorative techniques as a traditional rhetoric, and the tactile qualities all come together. While concentrating all possible attention to detail, it is at the same time a pursuit for a multi-layered hybrid expression.

UTA (Gachi Series)
1985
Yamaha Corp.

KAORI (Gachi Series)
1986
Yamaha Corp.

TAE (Gachi Series)
1986
Yamaha Corp.

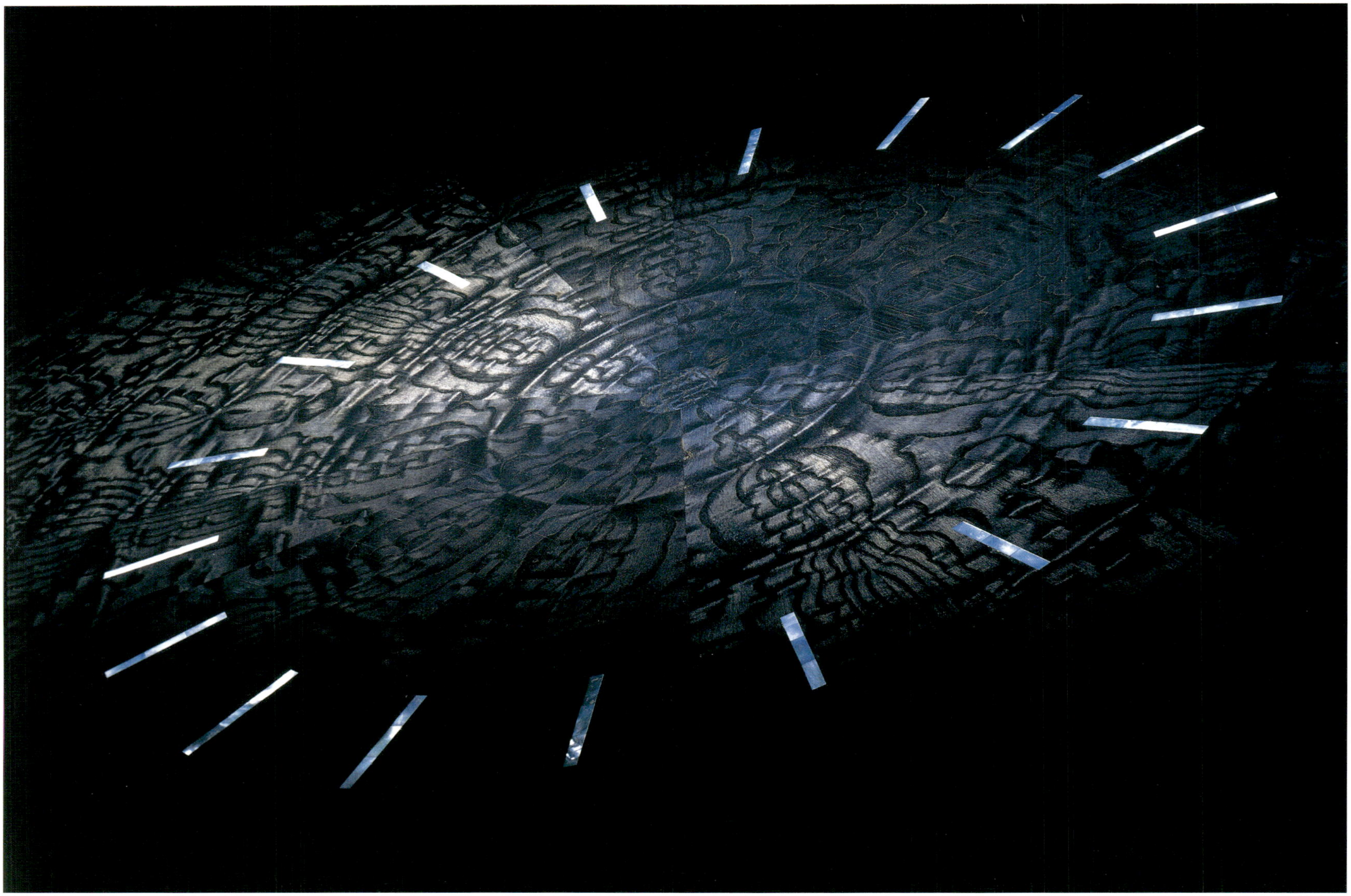

ROU (Gachi Series)
1986
Yamaha Corp.

HANA (Gachi Series)
1985
Yamaha Corp.

BRONX 1010 (Folding Chair)
1992
Cassina-Interdecor Japan Inc.

Folding Chair "BLITZ (TUNE)"

After winning first prize in the A.I.A. International Chair Competition in 1977, it took five years for this chair to go on the production line. It is not unusual for an industrial product involving dies for press or injection to take some time before mass production, but in this case the change of manufacturer caused further delays. The time was shortly after the oil crisis when prices of petrochemical materials surged, and the development from the conceptual design to the production model sought after a comfortable yet slim chair with minimal use of material.

BLITZ (Folding Chair) TUNE
1981-94
Skipper s.p.a.
1995- Cassina-Interdecor Japan Inc.

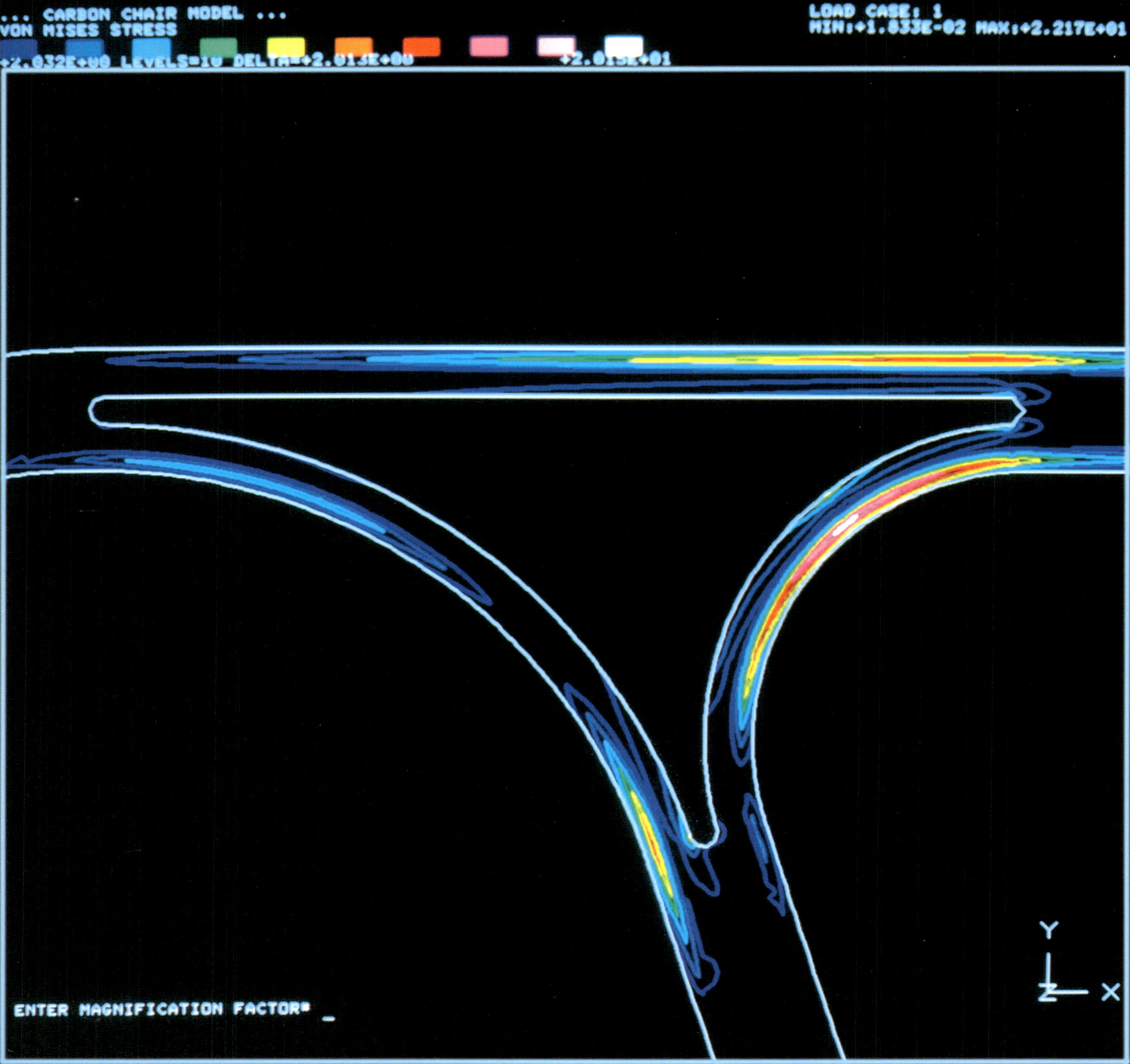
... CARBON CHAIR MODEL ...
VON MISES STRESS
LOAD CASE: 1
MIN:+1.833E-02 MAX:+2.217E+01
+2.032E+00 LEVELS=10 DELTA=+2.013E+00
+2.015E+01
ENTER MAGNIFICATION FACTOR= _
Y
Z
X

CCC (Carbon Composite Chair)
1985-90
Yamaha Corp.
1994- Okamura Corp.

ASCEND (Stacking Chair)
1987
Houtoku Co.,Ltd. Artifort

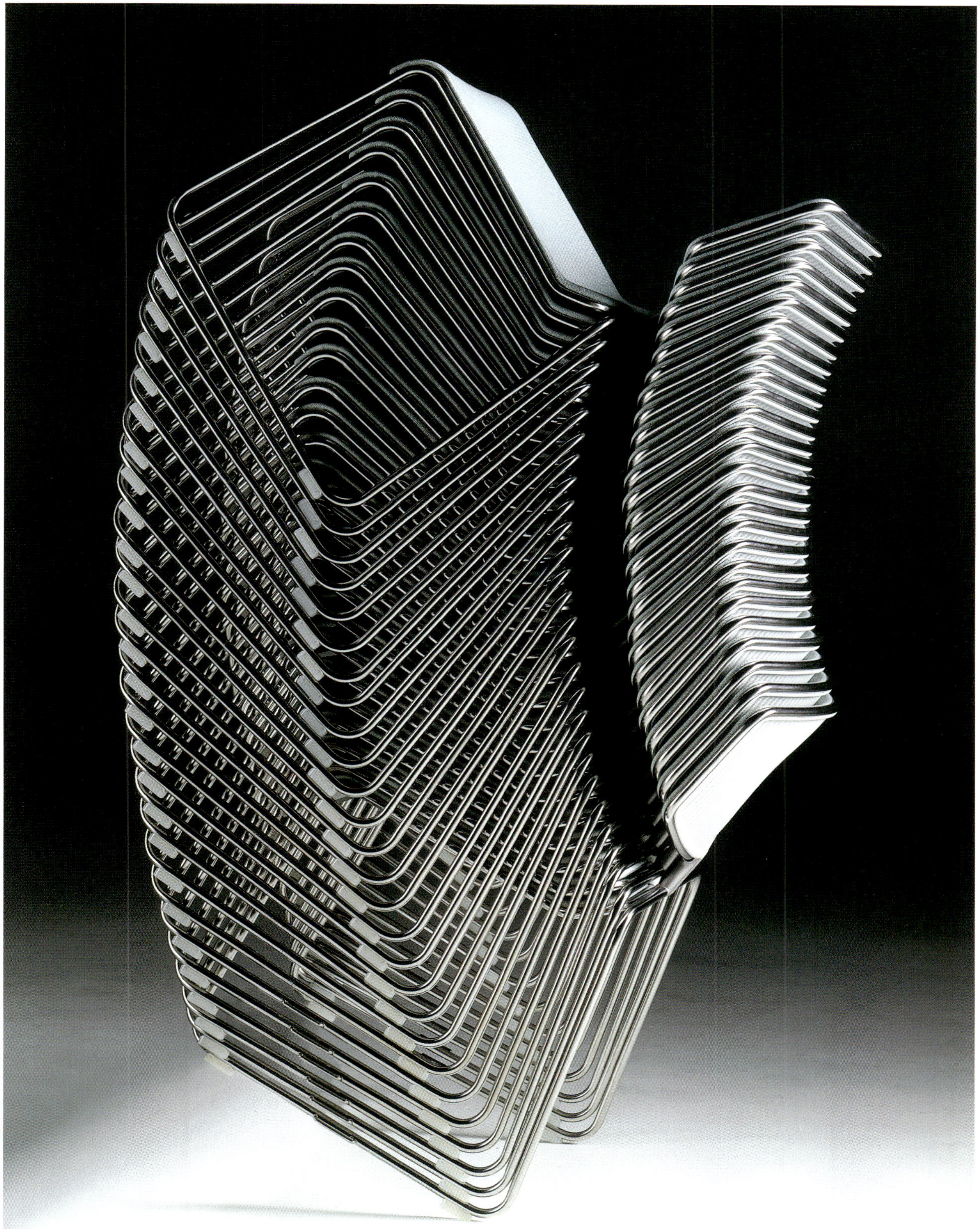

ACTON
2000
Iyobe Kogeisha Co.,Ltd. Okamura Corp.

FLESSO
2000
Iyobe Kogeisha Co.,Ltd.

ECHO
1991
Minerva Co.,Ltd.

CARBON CHAIR
(Prototype)
1988

ARIA SERIES
1992
Iyobe Kogeisha Co.,Ltd.
Sedie Friuli s.r.l.

JACKSONS
1986
Takashimaya Kosakusho Co.,Ltd.

SHIBA PARK TOWER / ENTRANCE HALL
2001
Mitsui Real Estate / Mitsui Mansion Project

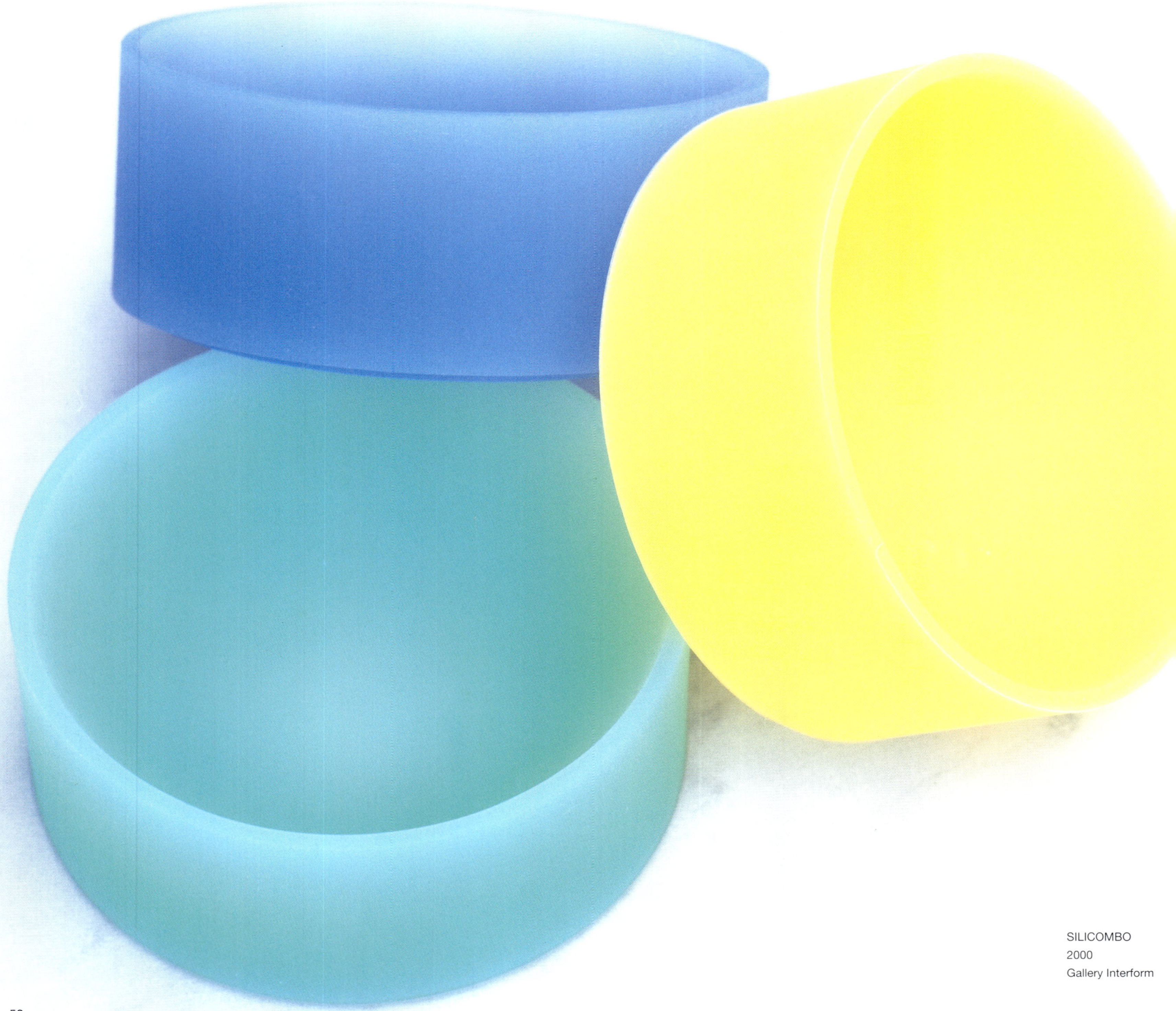

SILICOMBO
2000
Gallery Interform

TITAN (Tea Ceremony Spoon)
1993
Takenaka Seisakusho Co.,Ltd

Toward a Harmony of Living Spaces and Artifacts – From Furniture to Appliances

The intertwining of continuity and discontinuity from the past is a characteristic of today's Japanese culture, and it has its reflection in everyday living spaces. With the diversification of personal values and lifestyles, the number of artifacts surrounding us has exploded, adding further chaos. It is no longer an easy task to avoid drowning in quantity while pursuing true quality.

Interior design has always worked with furniture and decoration, but it is about time more attention was paid to domestic facilities and appliances as elements which support indispensable modern convenience and living. Information technology appliances deserve special attention, with their existence in homes quickly growing. The concept of this TV set was devised to answer future needs for larger displays, coordinating multiple information services such as security systems, on-line medical services and networking home appliances. The intention of this large but thin display was to avoid overwhelming the surroundings, both physically and mentally. Providing high quality in both function and design, this innovative interface, an "information window", is a proposal for a new life style.

PLASMA TELEVISION PX-42, PX-50
1998
NEC Corp.

PLASMA TELEVISION　PX-42, PX-50
1998
NEC Corp.

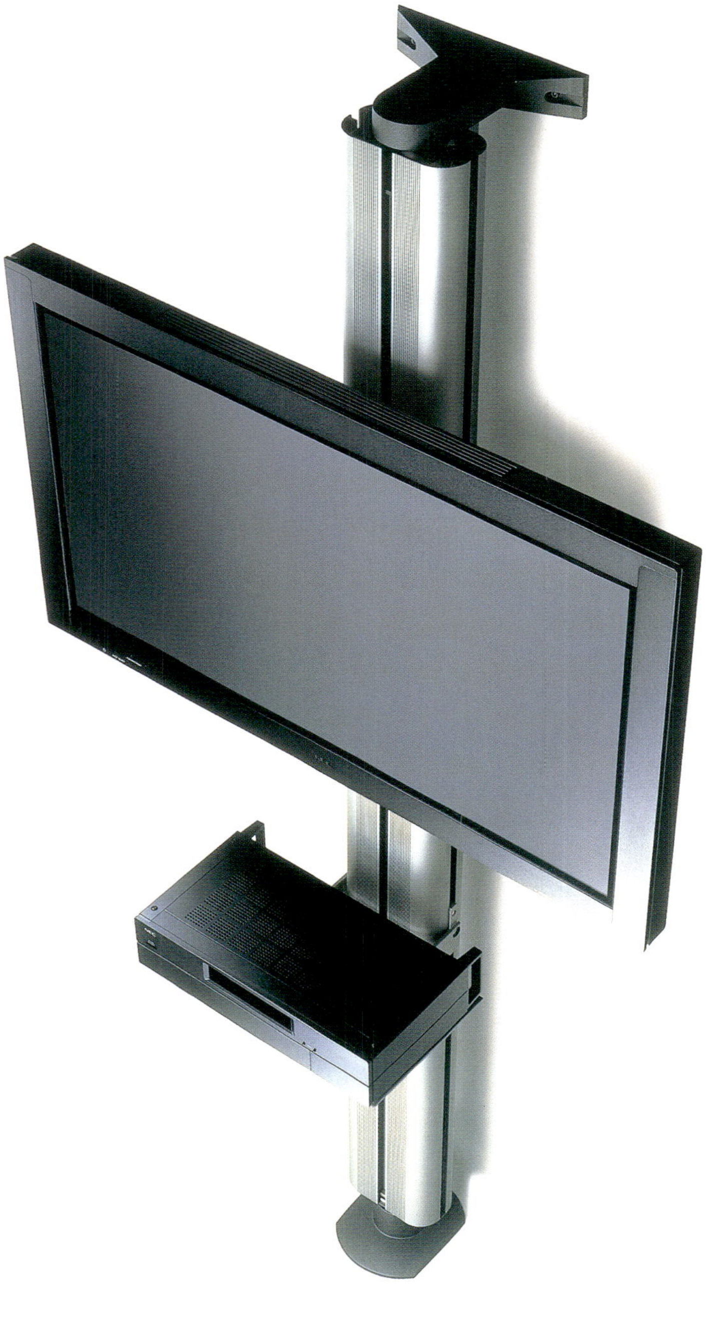

NEO MIND GEAR-DRESSER
1995
Shizuoka Furniture
Cooperative Association

MUG CUP
1992
MTG Design

BLENDY (Instant Coffee)
1986
Ajinomoto General Foods Inc.

MARIM (Powdered Cream)
1987
Ajinomoto General Foods Inc.

MALAY
1988
Central Market Malaysia
Central Market Japan Co.,Ltd.

GLASS TRAY
1975
Tachikichi Co., Adam & Eve

NEWS (Whisky Bottle)
1983
Kirin-Seagram Ltd.

KAGA STICK GREEN TEA
1991
Maruhachi Seichajo Ltd.

Japanese Green Tea House ISSYO　Kanazawa

TOKYO XII
1991
MTG Design

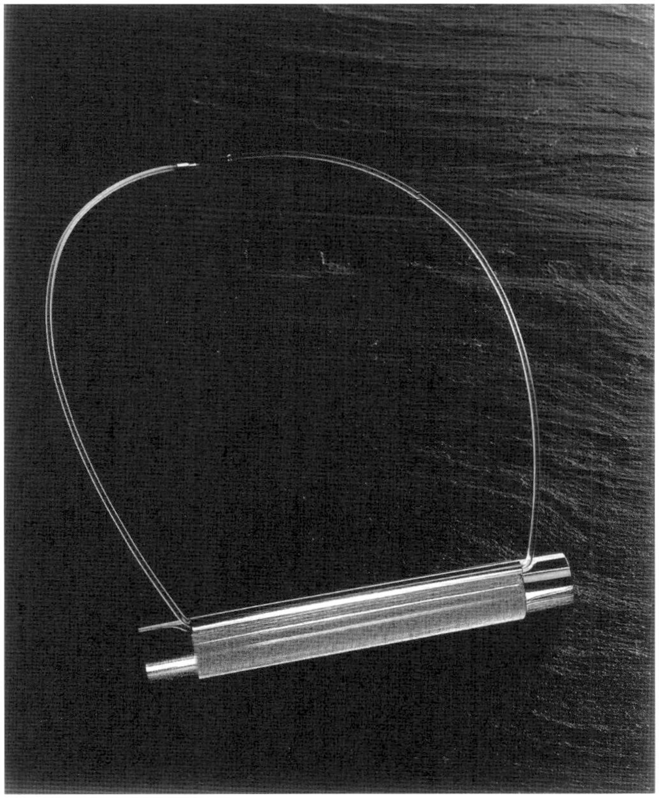
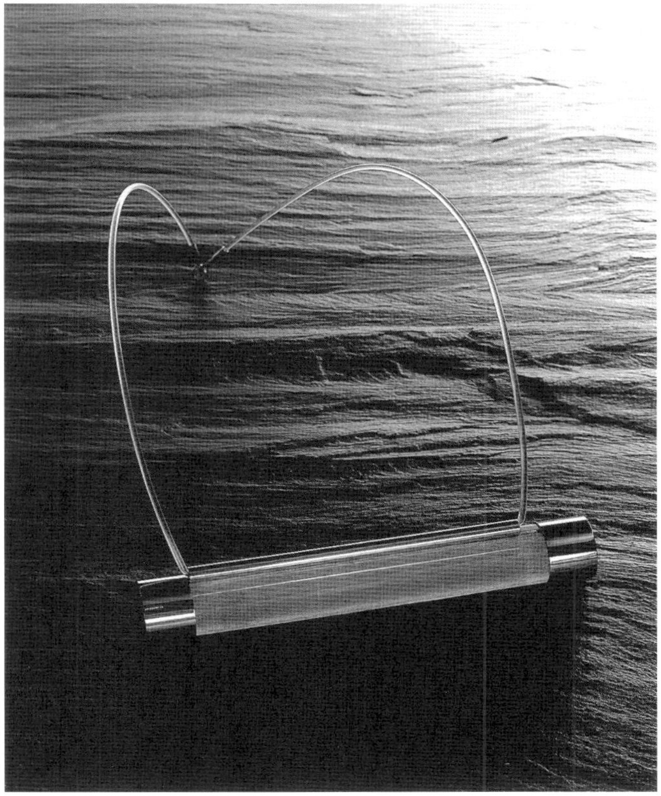

METALMESH NECKLACE
1983
Limited Edition

Harmony of Light and Shade

People naturally look for bright lights in spaces for activity and softer lights for relaxing. A combination of light and shade suitable for each place and situation is indispensable for a comfortable home interior.

With the fall of dusk, artificial lighting produces the atmospheric effects for conversation over dinner or relaxation in the living room. Whether letting the light reflect on the ceiling or wall to produce a soft and soothing effect or creating a pool of light at the sofa for relaxation, I always try to avoid a bright and monotonous light, and design fixtures that utilize various techniques of direct and indirect lighting.

GEN SERIES (Bracket)
1988
Yamagiwa Corp.

RUACUL
1978
Yamagiwa Corp.

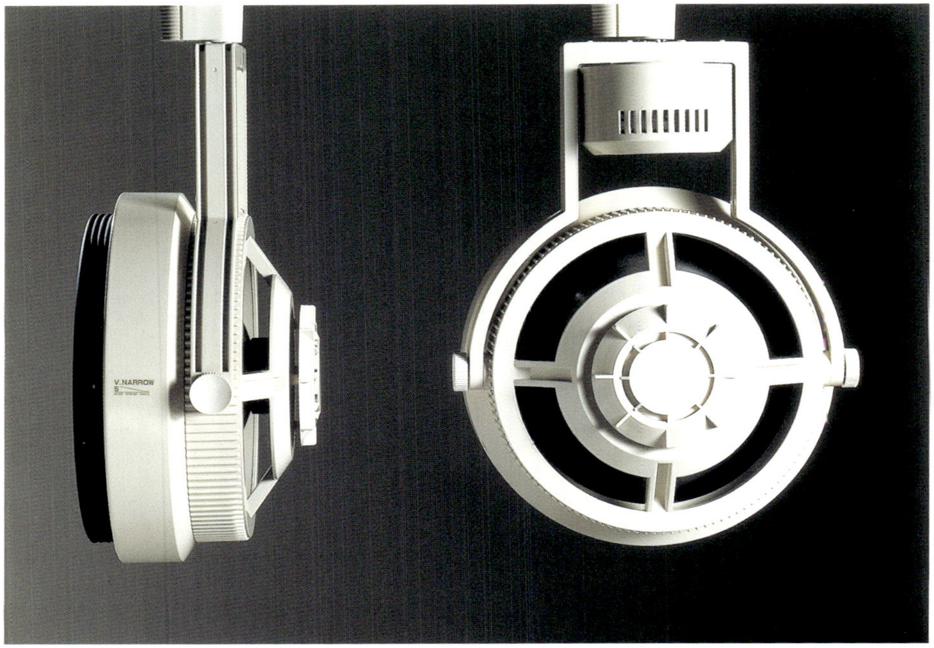

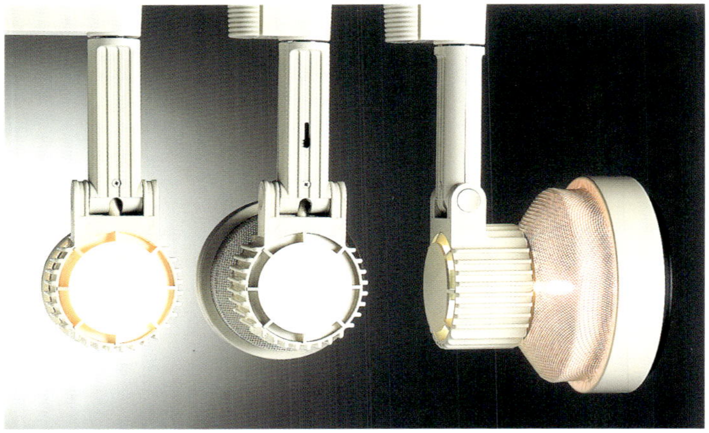

ML SPOT LIGHT
SUPER,HYPER 1988
Yamagiwa Corp.

RUACUL
1978
Yamagiwa Corp.

ARIAKE (Floor Stand Lamp)
1983
Bushy Co.,Ltd.

MODERN-JU
1981
Bushy Co.,Ltd.

KASANE
1982
Bushy Co.,Ltd.

KINU SERIES (Cabinet)
1982
Arflex s.p.a.

KINU SERIES (Chair)
1982
Arflex s.p.a.

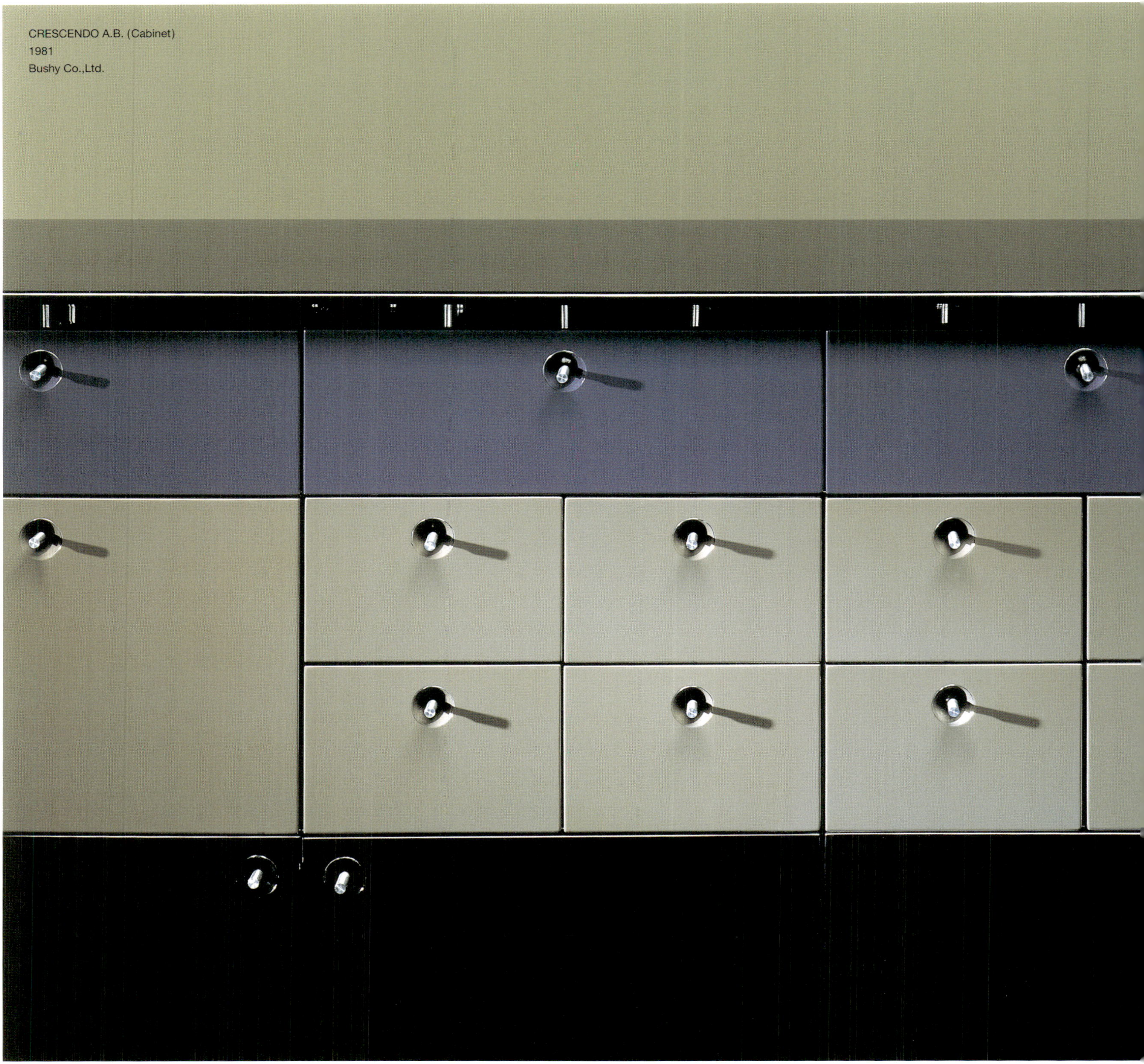
CRESCENDO A.B. (Cabinet)
1981
Bushy Co.,Ltd.

SANITARY ACCESSORY MIRROR (MK Series)
1996
TOTO Co.,Ltd.

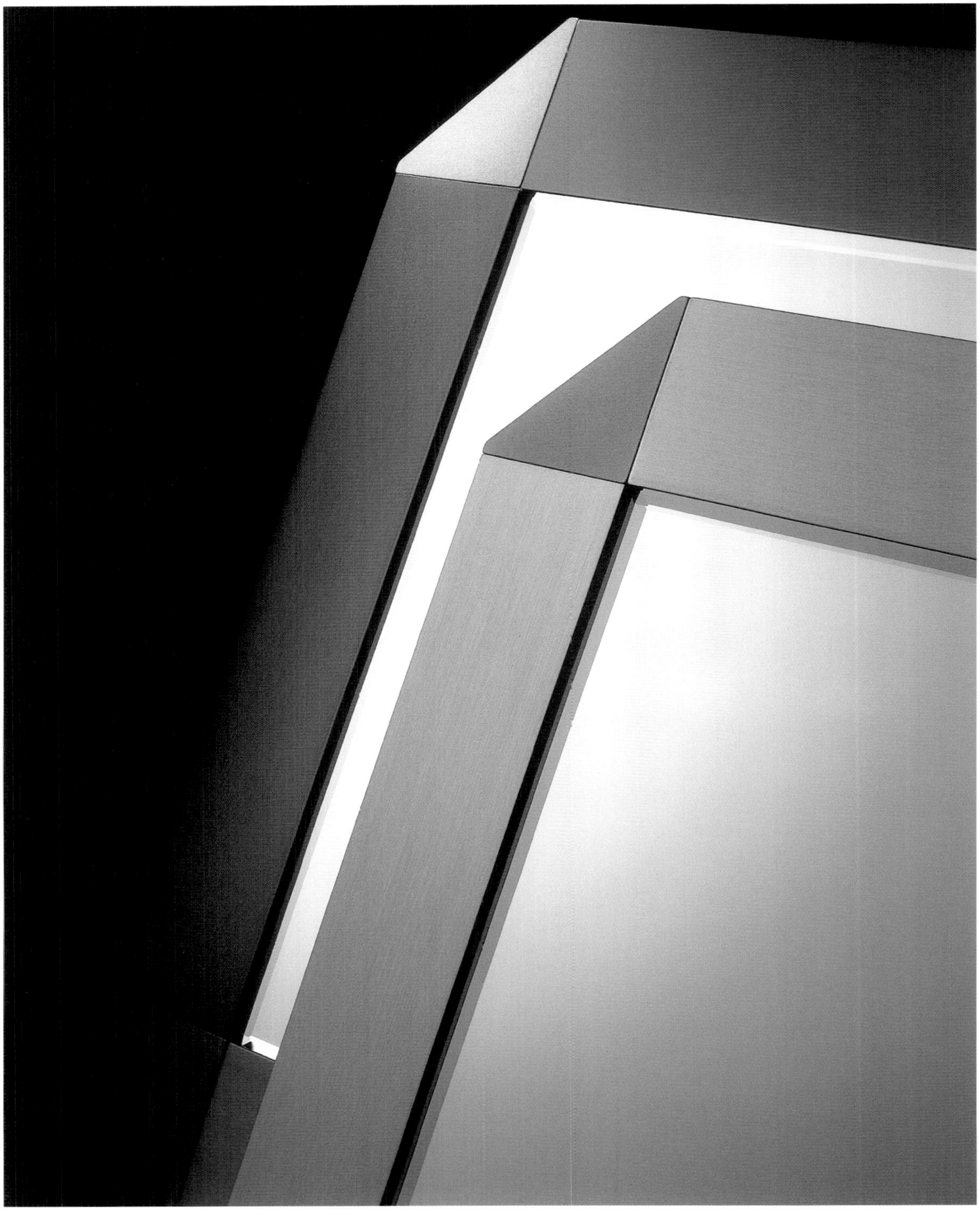

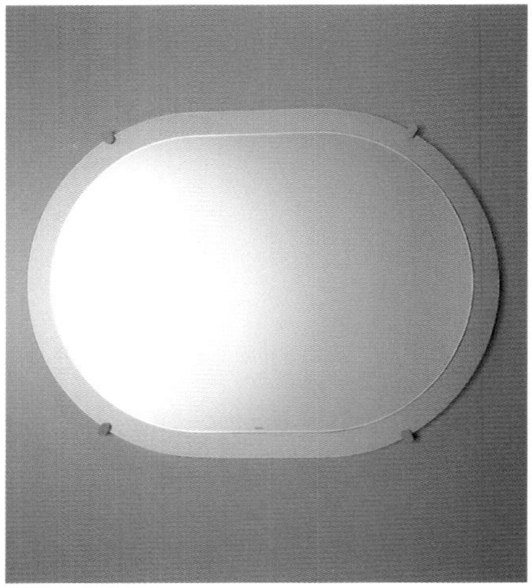

PUSH & PULL
1974
Arflex Japan Ltd.

The Tsurumi Tsubasa Bridge

Today we designers have more chances than before to participate in landscape projects such as public architecture, bridges, roads, parks, and riverside developments, resulting in more opportunities to cooperate in making beautiful landscapes and better living environments for the next generation.

The "Tsurumi Tsubasa Bridge" was one project I was involved in for six years as a landscape designer. It is one of the main bridges of the Shuto Expressway Wangan Line, connecting Haneda Airport and Yokohama Bay Bridge. A diagonal suspension bridge with a span of 1,020 meters, the bridge is surrounded by typical industrial docks, while the waterway beneath is frequented by container ships of the 50,000-ton class. Considering this busy and chaotic environment, I looked for a harmony and beauty suitable for the surroundings, aiming to achieve a symbolism and a calm visual effect at the same time.

TSURUMI TSUBASA BRIDGE
1994
Metropolitan Expressway Public Corporation

Tokyo
Tokyo-Bay
Haneda Airport
Kawasaki
Yokohama
Tsurumi Tsubasa Bridge
©MIT-NASDA JERS-1

Tokyo
Tokyo-Bay
Haneda Airport
Kawasaki
Tsurumi Tsubasa Bridge
Yokohama
©MIT-NASDA JERS-1

NEX SERIES DESK
1999
Uchida Yoko Co.,Ltd.

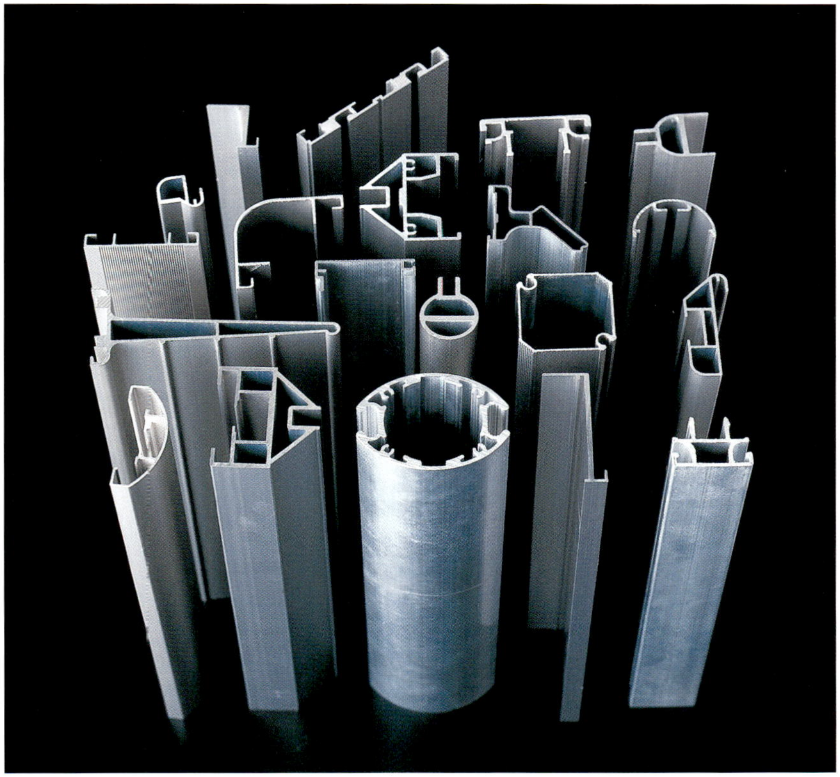

NEX SERIES ALMINUM FRAME

NEX SERIES PARTITION
1999
Uchida Yoko Co.,Ltd.

NEX SERIES
PARTITION 1999
Uchida Yoko Co.,Ltd.

LOBBY CHAIR FOR OZONE
1994
Okamura Corp.

NIKI CLUB NASU PART 1
1986
Yamaha Corp. Interior Center Co.

JAPAN NATIONAL CIL CORPORATION
TECHNOLOGY RESEARCH CENTER
1988
Houtoku Co.,Ltd.　Terdo Co.,Ltd.

RACK A,B (For CD)
2001
S.B.Decola & Building Materials Co.,Ltd.

SYSTEM KITCHEN SUPER LEGACESS-L
2001
TOTO Co.,Ltd.

SYSTEM BATHROOM
(Furopia KV Series) 2001
TOTO Co.,Ltd.

ZAZA (Motomi Kawakami Collection)
1993
TOTO Co.,Ltd.

ZAZA (Motomi Kawakami Collection)
1993
TOTO Co.,Ltd.

SANITARY ACCESSORY (MK Series)
1996
TOTO Co.,Ltd.

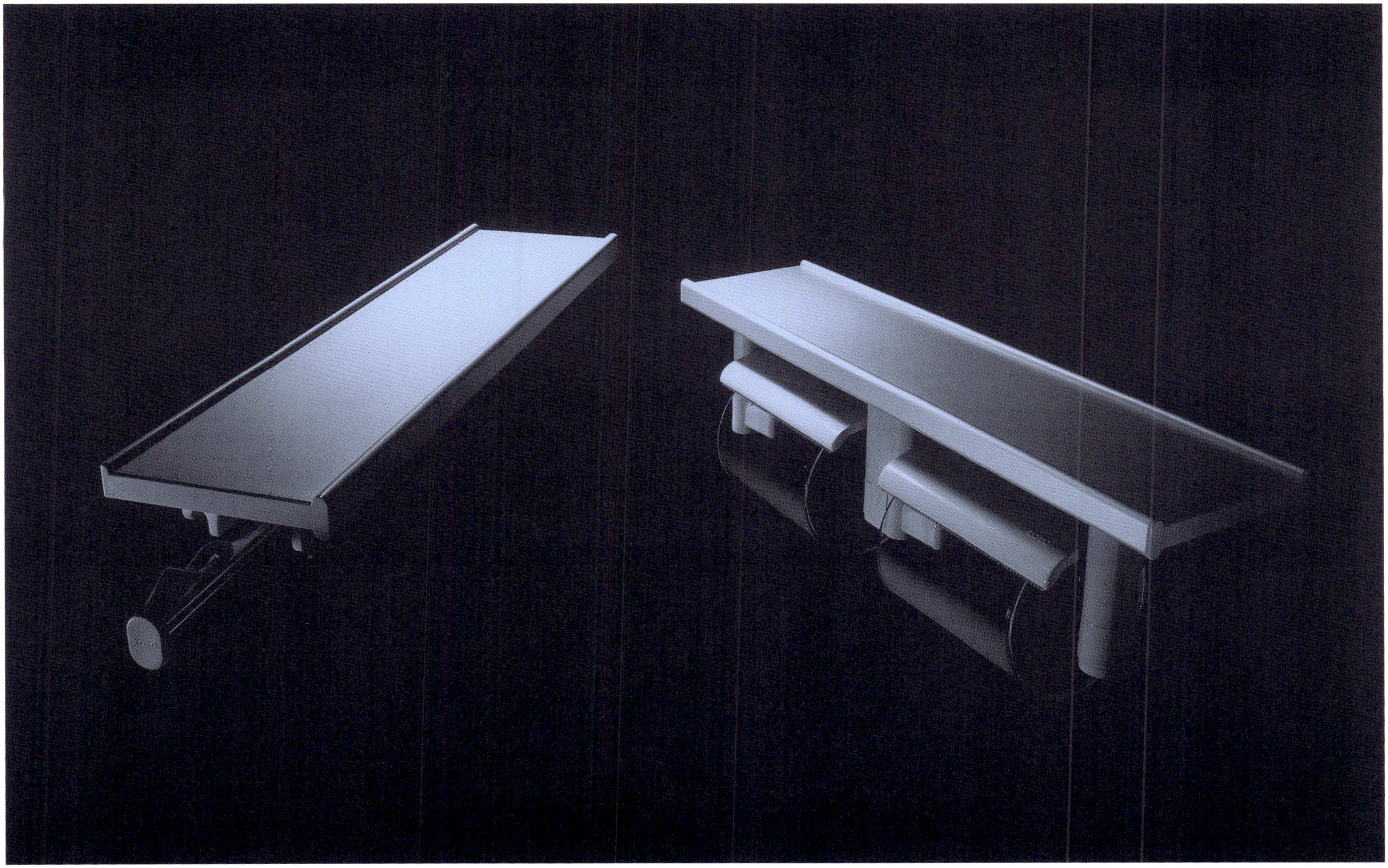

WASHBASIN PROJECT (Scale Model)
1994

SANITARY ACCESSORY (Material Design Series)
Glass, Aluminum, Wood
1999
TOTO Co.,Ltd.

PEDESTRIAN FURNITURE (Project) 1995

Walking is one of the basic activities in our life. The environments for walkways should have a friendly atmosphere and support intimate gatherings and rest. But in reality the streets, facilities, and roadside plants are too often either superficial decoration or simply too poor.

The existing situation consists of many elements layered one over the other, but here is one more attempt to bring some harmonious order into this environment, through the utilization of ready-made systematized units.

Based on a 60-centimeter grid system, these simple street furniture elements of various shapes are made of materials such as mock stone, cement and water-permeable pavement in different colors. This unit system is not intended to cover the whole landscape but rather to bring out the charm of each surrounding environment to form the basis of a warm landscape. A proposal for the key to "an environment for happy and lively walks."

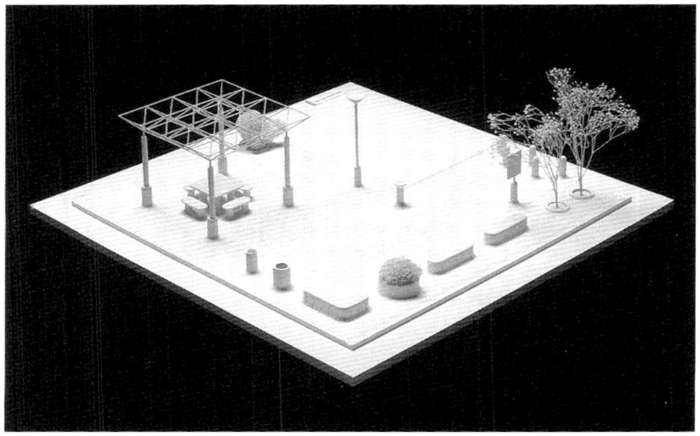

PEDESTRIAN FURNITURE (Project)
1995

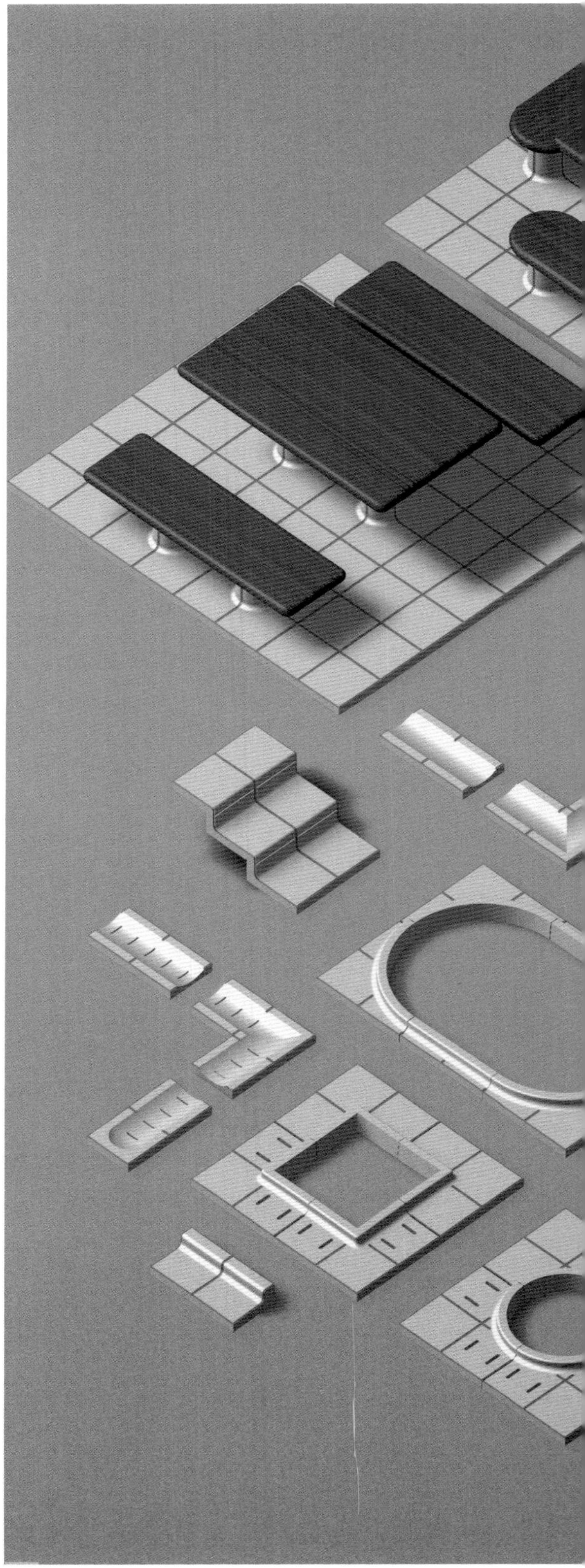

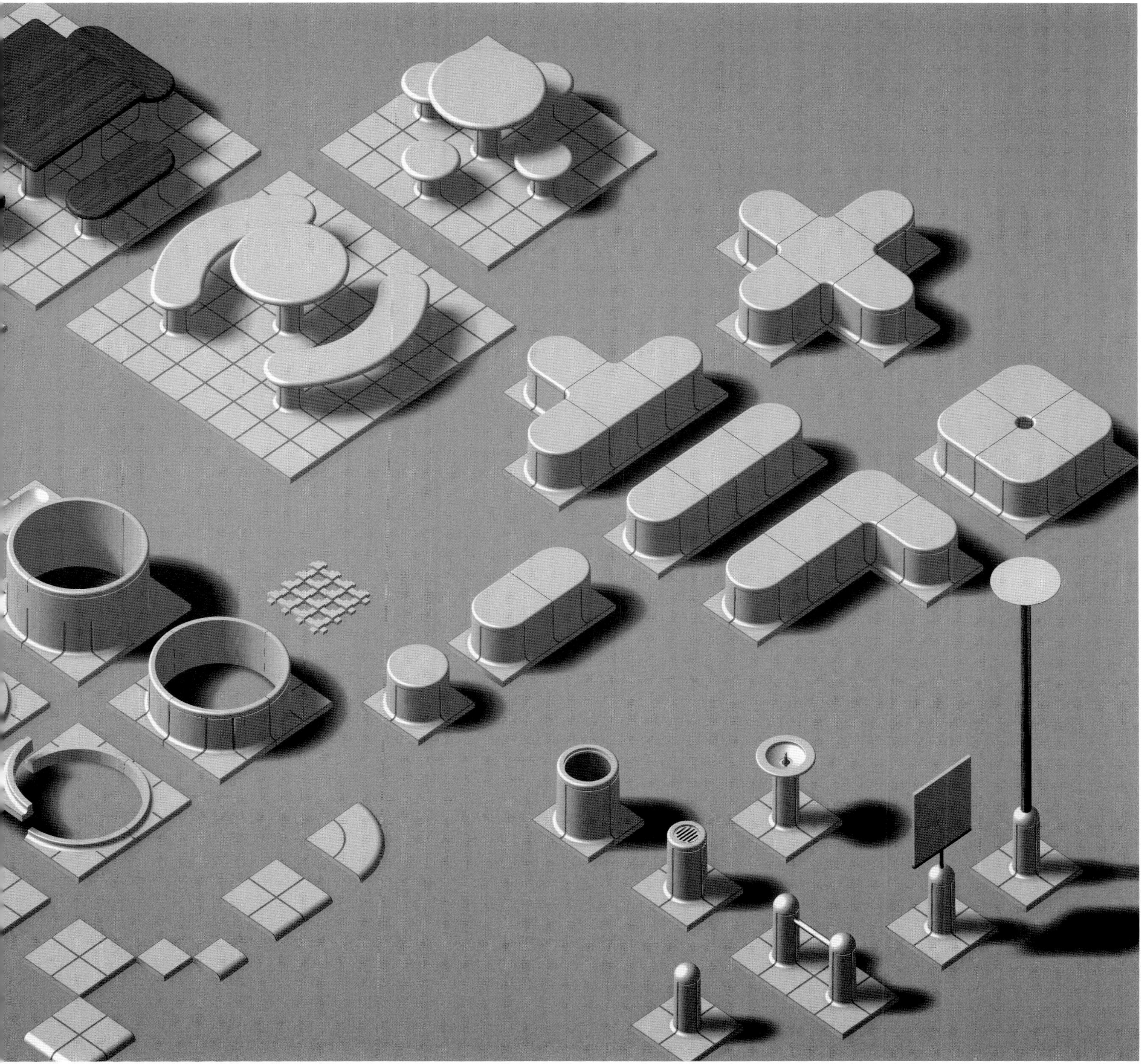

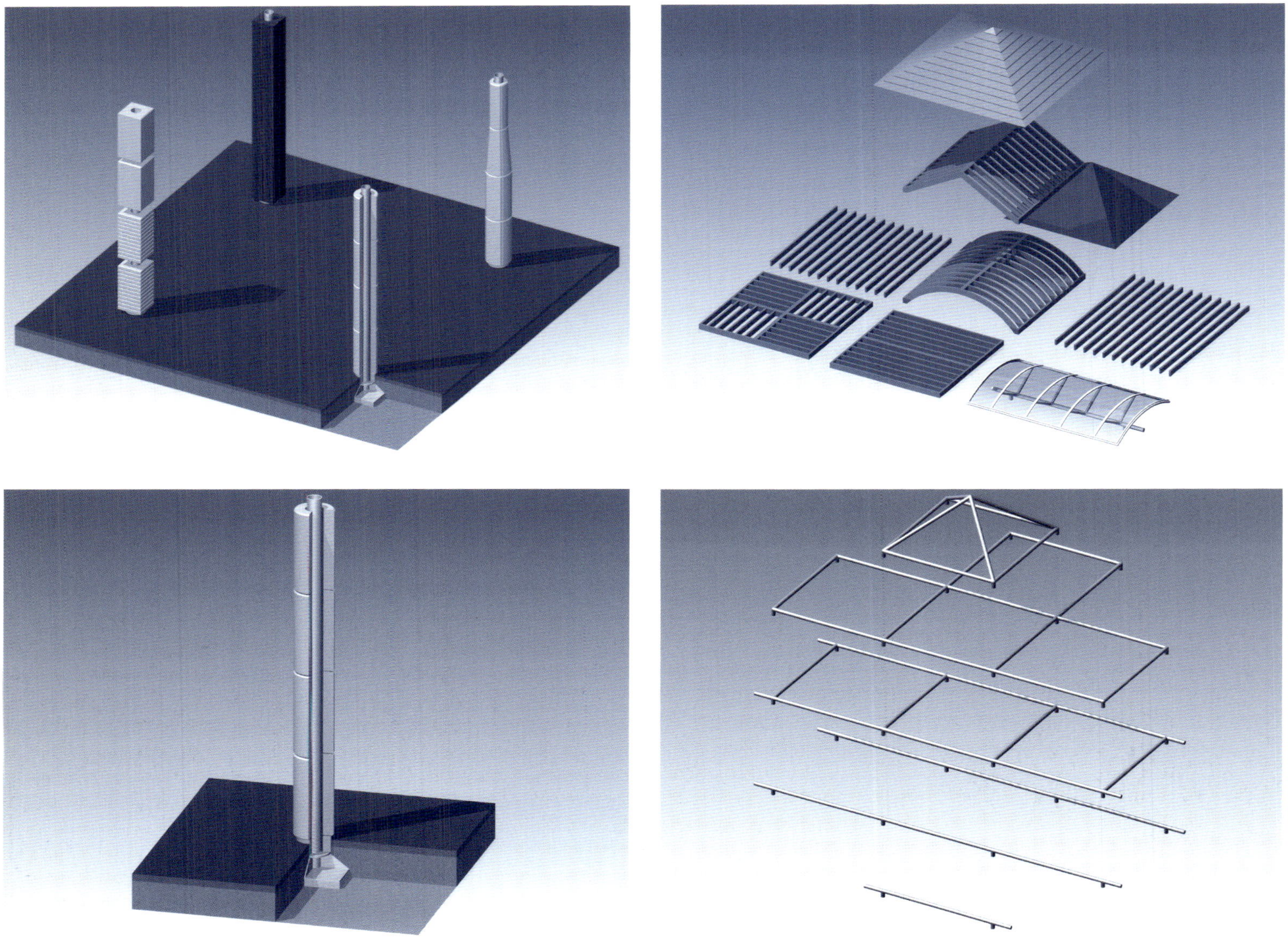

FERGOLAS & SHELTERS (KD Series)
1994
Hokusho Co.,Ltd.

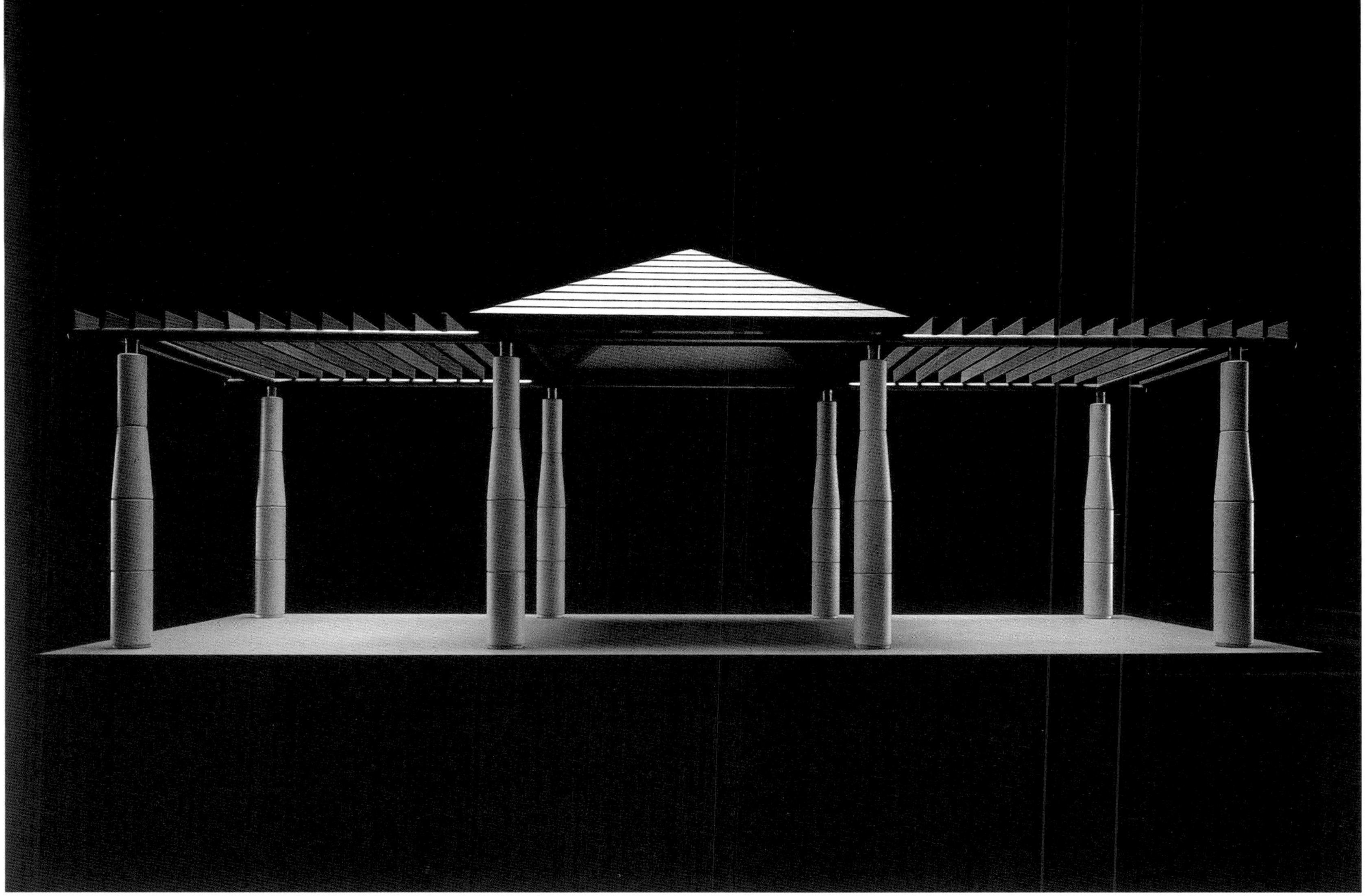

TENNIS RACKET (RQ Series)
1988
Yonex Co.,Ltd.

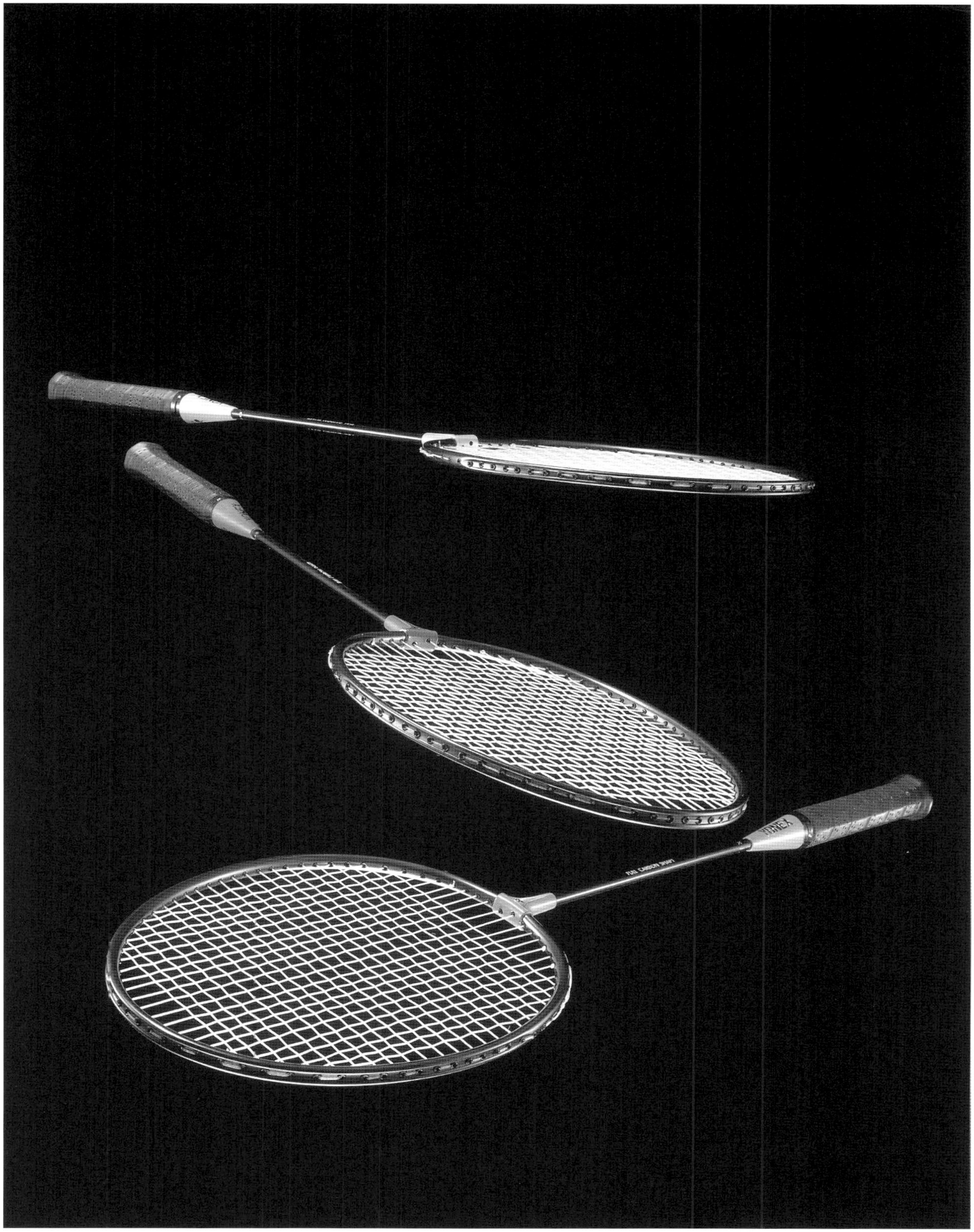

BADMINTON RACKET (Carbonex-8)
1986
Yonex Co.,Ltd.

INTERFLEX (Desk clock)
2000
Gallery Interform

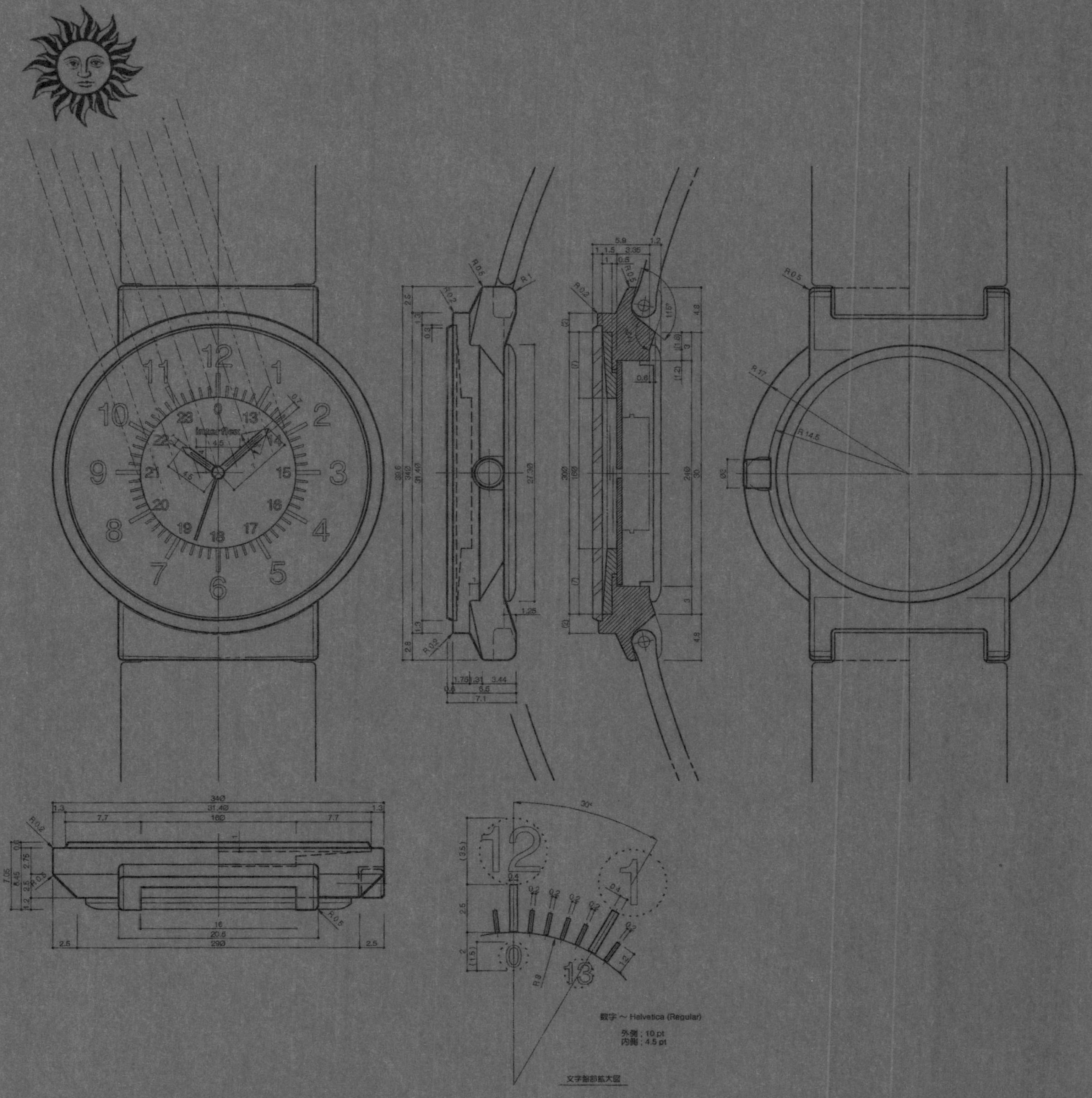

数字 ～ Helvetica (Regular)
外側 : 10 pt
内側 : 4.5 pt
文字盤部拡大図

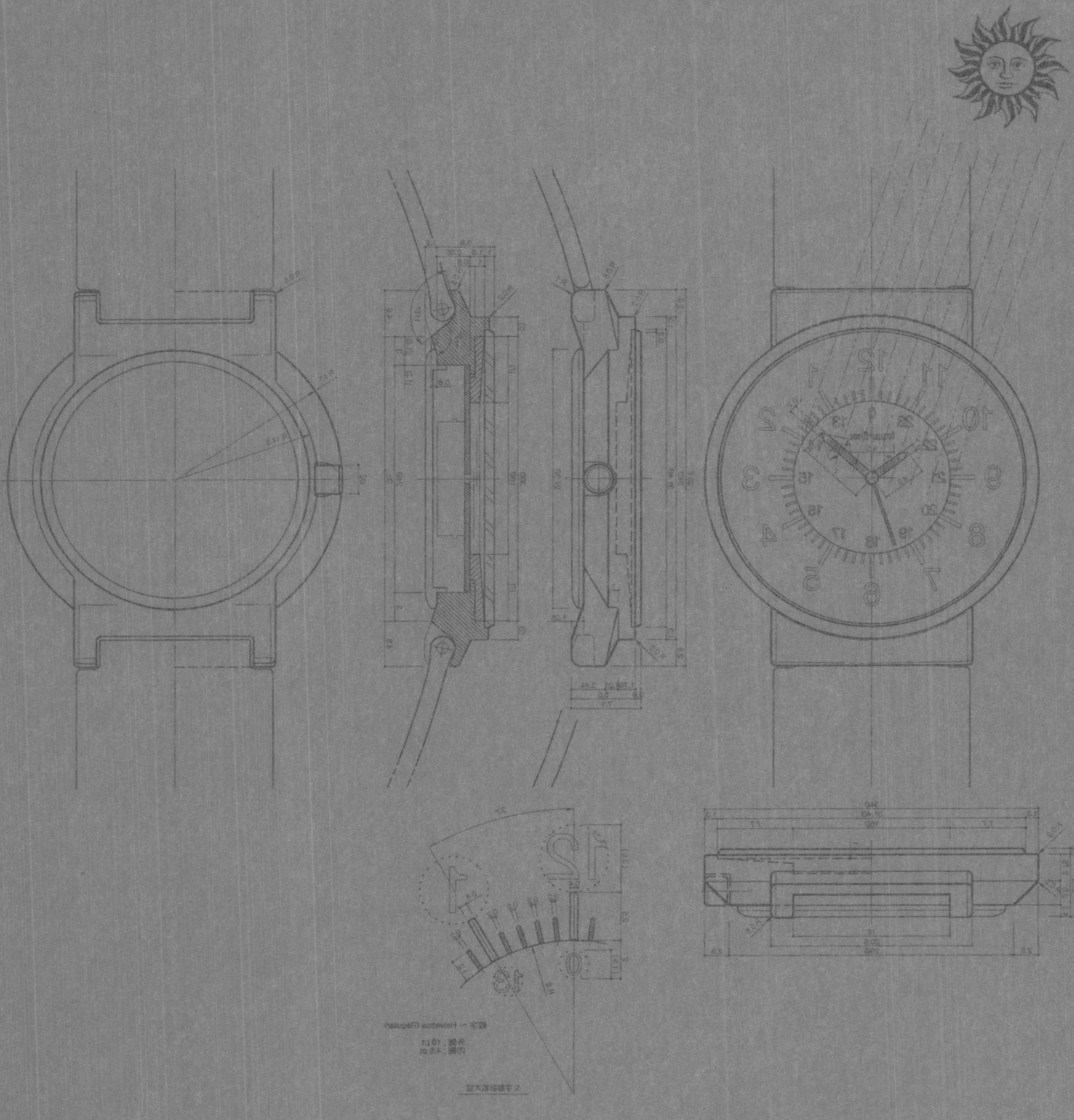

INTERFLEX (Wrist watch,Eco crive)
2000
Gallery Interform

INTERFLEX (Desk clock,Prototype-granite)
2001
Gallery Interform

INTERFLEX (Wall clock)
2000
Gallery Interform

HAND SCANNER
1976
Sumitomo Electric Co.,Ltd.

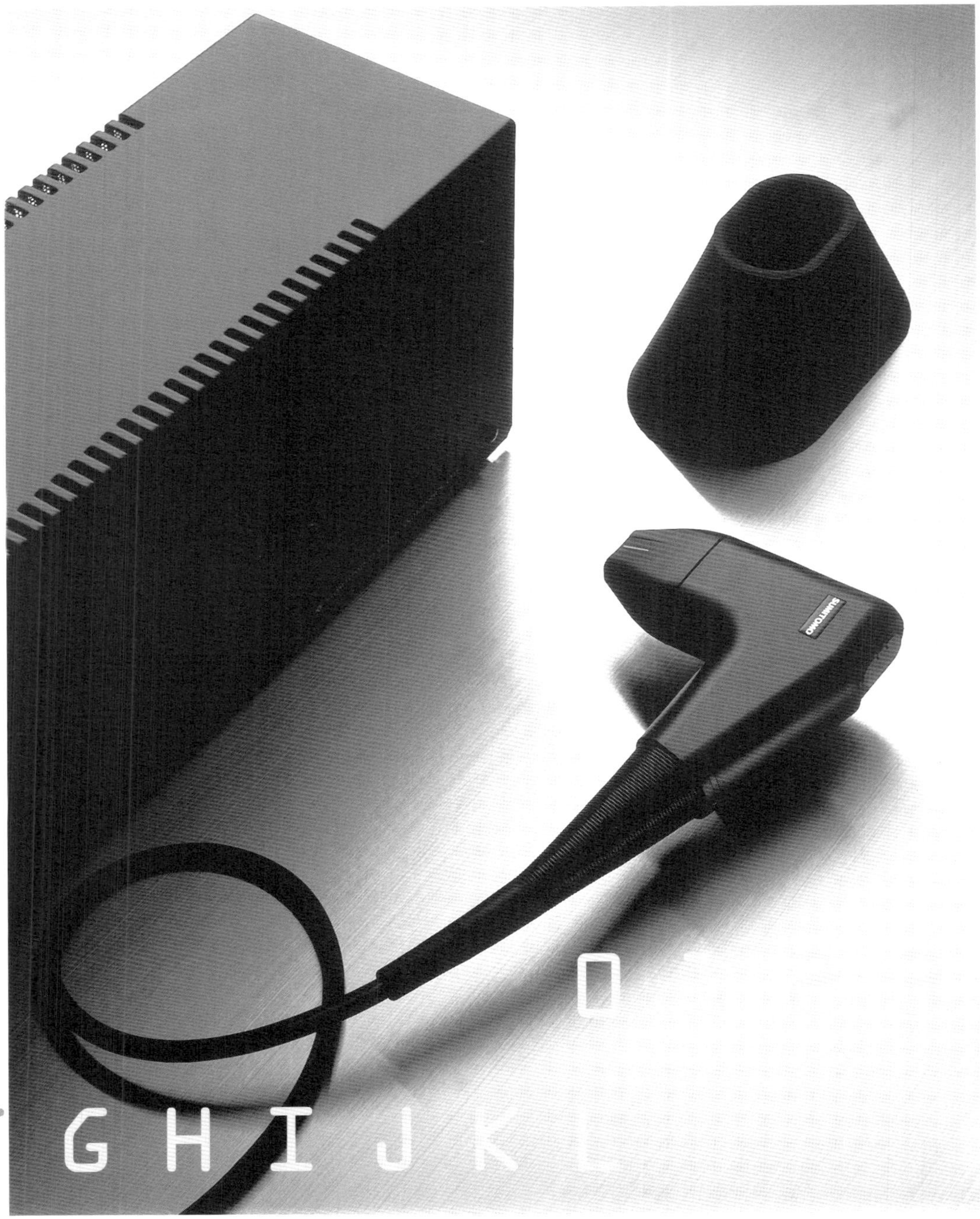

Motomi Kawakami Biography

1940

Born in Hyogo Pref.

1964

Graduated from Tokyo National University of Fine Arts and Music, majored in industrial design

1966

Graduated from the Masters course of the same university

1966-69

Professional practice at Angelo Mangiarotti Architect Office, Milan, Italy

1971

Established Kawakami Design Room
Design work specializing in product, interior, space and environmental design.
Collaboration on local industry promotion projects and professional education programs in Takayama, Asahikawa, Takaoka, and others

●Exhibitions, lectures, etc.

1969

"Modern Chairs 1918-1970" White Chapel Art Gallery's Exhibition (London)

1970

Good Design Exhibition "Global Eye" (Ginza Matsuya, Tokyo)

1971

Good Design Exhibition "My Seat" (Ginza Matsuya, Tokyo)

1978

"The Form of a Chair – From Design to Art" (The National Museum of Art,Osaka)

1982

Joint Exhibition in Five Galleries (Furniture)
Gallery LUFT (Tokyo),
Bushy showroom (AXIS, Tokyo),
Skipper showroom (Milan),
Arflex showroom (Milan),
Design 19 (Ginza Matsuya, Tokyo)

1984

"Design in Japan – Tradition and the Present" exhibition organization and installation (Moskow)

1987

One-man furniture exhibition "Hybrid–Light and Shade" (Gallery MA, Tokyo)
"Contemporary Scandinavian Crafts" exhibition coordination and installation (Yurakucho Seibu Art Forum, Tokyo / The National Museum of Modern Art, Kyoto)

1988

One-man exhibition "Fragments of Form – Furniture and Lighting" (Creators Gallery, Yurakucho Seibu, Tokyo)
Symposium "Designers Week-End" panel speaker (Brussels)

1991

One-man furniture exhibition "Industrial Symphony" (Quest Hall, Tokyo)

1995

"Japanese Design-A Survey Since 1950" (Philadelphia Museum of Art)

1996

Wood of Finland "Innovative Industrial Wooden Products" competition juror and lecturer (Finland)

1997

Ministry of Trade and Industry "Good Design Award" head of general jury

1999

"New Perspectives on Industrial Design" exhibition coordination
(The Museum of Modern Art, Toyama)

2000

"Masterpiece Chairs of the 20th Century – 200 Chairs by 200 Designers"
(Kobe Fashion Museum)
One-man exhibition "Material Message" (Ozone Gallery, Tokyo)

●Books and articles

'The Beauty of Bridges II : Road Bridges Landscape Handbook' 1981, Japan Road Association
'GACHI – Motomi Kawakami's Furniture' 1988, Rikuyo-sha
Article in 'Dictionary of Furniture' 1986, Asakura Shoten
Article in 'Dictionary of Design' 2000, Heibonsha Ltd., Publishers

●Honors

1967

Bronze Prize, Monza Furniture Competition (Concorso MIA), Italy

1968

14th Milan Triennale
Gold Prize, Geneva Furniture Exhibition

1972

Silver Prize, Yamagiwa International Lighting Design Competition

1976

Japan Interior Designers' Association Award

1977

1st Prize, The Open International Chair Design Competition (The American Institute of Architects)

1981, 82

IF Prize (West Germany)

1984

Prize for Excellence, Annual Japan Interior Design Award

1985, 86

Japan Package Design Award for Super Excellence (85), Excellence (86)

1991

Mainichi Design Prize

1992

Kitaro Kunii Industrial Design Award

1995

Tanaka Prize, Japan Society of Civil Engineers for "Tsurumi Tsubasa Bridge"

1996

Yokohama Civic Design Award

1998

Gold Prize, Good Design Award for NEC Plasma Display TV PX-42,
Prize for Excellence, *Wood Chairs in Living Exhibition*

1999

IF Prize (Germany)

OBJECT LIST

P8
FLOWER VASE
1999
Self product
W200×D50×H200mm

P9
WOOD SAMPLING 91
1999
Gallery 91(New York)
W220×D200×H190mm

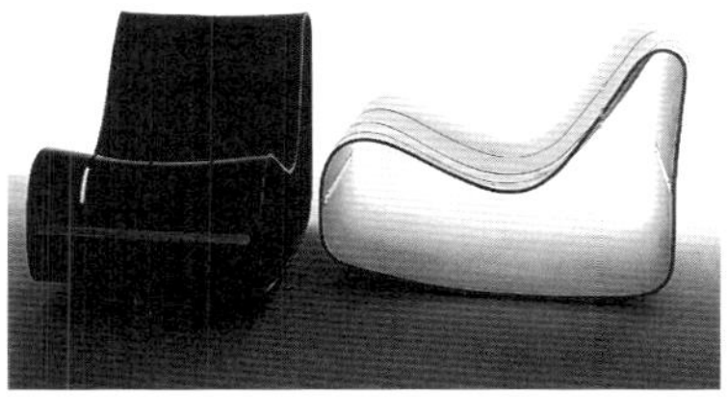

P10
FIORENZA
1968
Alberto Bazzani s.p.a.
W570×D900×H660mm

It was firstly commercialized in molded plywood. As it was right in the middle of the plastic age, and moreover there were huge needs around the waterfronts, it was altered to heat molded plastic type. Exhibited in the Triennale, Eurodomus and others.

P12,14,15
NT (Chair)
1976
Arflex Japan Ltd.
W556×D565×H770 SH440mm
Stand chair: W489×D509×H977 SH695mm
Baby chair: W450×D470×H777 SH495mm

As a dining chair, a personal chair in the study, or a public chair in meeting rooms or halls, this chair has various usages and gives timeless comfort. Made of plywood frame and woven tape seat exploiting their characters. Simple frame draws curves which fit human bodies gracefully, and the seat and backrest woven with tapes offer moderate cushion. In 1995 was presented NT SLIM which was a little more compact and light.

P13
NT Poster

P16,17
MK (Chair)
1992
Arflex Japan Ltd.
W490-565×D530×H785 SH440mm

Making the joints bigger it gains lightness and stability, and elastic fabric provides seating comfort.

P18,19
LUCA (Chair)
1997
Arflex Japan Ltd.
W502-530×D515×H780 SH440mm

The backrest is woven in a mesh made of 2.4mm-thick strips of wood molded separately. The frame is made of beech wood finished with polyurethane lacquer, and the armrests are the combination of cast aluminum and bent beech wood.

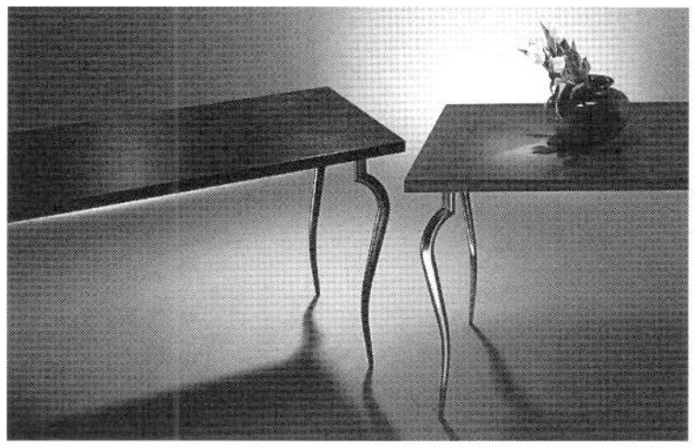

P20
ASHI (Table)
1992
Acerbis International s.p.a.
W1400-1800×D800×H720mm
W1000-1200×D1000-1200×H720mm

P21
MUSA (Chair)
1991
Acerbis International s.p.a.
W480×D520×H800mm

At the very dawn of human civilization most seating was highly decorated as a symbol of authority. But from the time of ancient Greeks, there is evidence too in a number of reliefs, of the existence of chairs in daily use known as "Klismos", used mainly by the ladies. Klismos had a simple shape, elegant, always comfortable and ideal for everyday usage.
These origins and this spirit, the human shape itself, made a strong impression on me and I designed "Musa", the chair for contemporary use. Musa was created as a homage to Klismos.

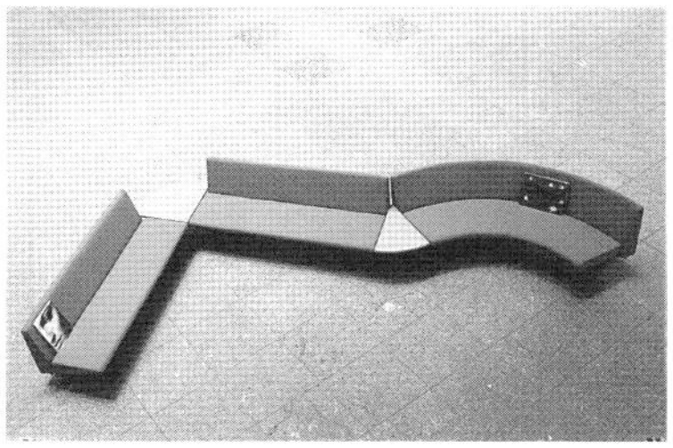

P22, 23
TICINO - Sopra Fiume Series
1991
Okamura Corp.
W1800-2120×D650×H720 SH380mm

In city halls, hotels, banks, galleries or hospitals... These reception seatings are, with the state of congested reception spaces in mind, named "SOPRA FIUME" after the manner and flow of the people gathered and chatting, reminiscent of the current of rivers.
As the embodiment of human activities in different places, I designed these seatings to be one vision of space embodied in industrial facture. TICINO has connecting benches and corner tables resembling a meandering river in the plain, dividing halls and lobbies and creating rhythm in a monotonous space. Simple and flowing form.

P24
OMBRONE - Sopra Fiume Series
1991
Okamura Corp.
1991
W1968-2220, W1312-1566, W650-910×D857×H718-802 SH390mm

One seater, two seater, three seater, with armrests, without armrests, bench type, and a glass table. The backrest has a two-level angle adjuster, providing deep assurance and a sense of quality.

P25
OFANTO - Sopra Fiume Series
1991
Okamura Corp.
W1100-1880×D857-910×H736 SH380mm

The color and form give a sense of great volume despite its airiness like chatter of gill. There are a center table and a side table formed and textured like a leaf.

P26
PIAVE - Sopra Fiume Series
1991
Okamura Corp.
W800×D810-822×H760-1010 SH380mm

The existence adds accent to a space like a monomerous flower appearing out of the ripply water.

P26
VOLTURNO - Sopra Fiume Series
1991
Okamura Corp.
W860-2090×D770×H700　SH385mm

Solid, massive and rigid-formed chairs. Between humans and spaces, their existence is much more closely connected to spaces.

P27
PLATANI - Sopra Fiume Series
1991
Okamura Corp.
W1800×D610×H770-725　SH380mm

The legs have delicate details like rings of water. These light and smooth textured benches are formed with SKF frames in their armrests and legs.

P28, 29
METAURO - Sopra Fiume Series
1991
Okamura Corp.
W1840-3100×D580-630×H780　SH410mm

The joint type of 3 to 5 seater, the bench type and the center and side table. The backrests and seats have a rounded shape like stones washed and beaten in the river. Having solid aluminum texture as well as soft touch for the seating which can be reclined.

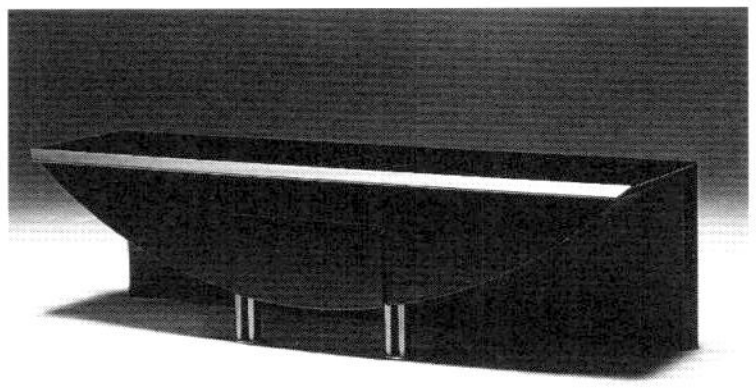

P30
UTA - Gachi Series　(Sideboard)
1985
Yamaha Corp.
W1605×D460×H855mm

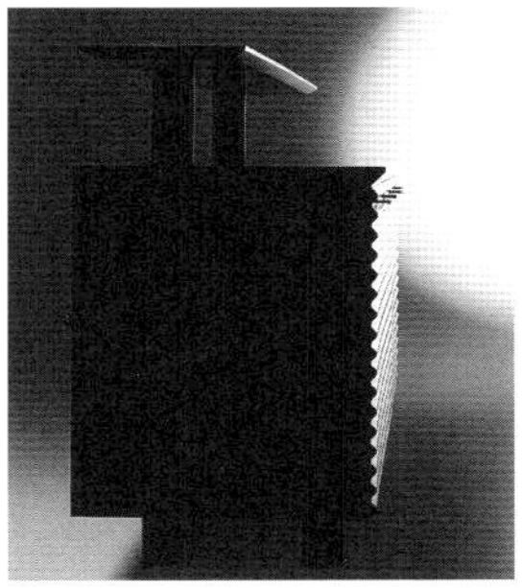

P30
KAORI - Gachi Series　(Sideboard)
1986
Yamaha Corp.
W1800×D450×H500-700-500mm

P31
TAE - Gachi Series　(Writing cabinet)
1986
Yamaha Corp.
W1200×D435×H1200mm

"Tokonoma" featured in traditional Japanese living space reflects the changes of seasons and expresses spirituality. Trying to exploit this mentality, I designed this cabinet as a piece of furniture in a contemporary living space.

P32
ROU - Gachi Series　(18-Cornered table)
1986
Yamaha Corp.
ϕ1500×H700mm

Crepe-patterned brown ash, pearl marquetry finished with mirror-surface coating of polyurethane. Huge wood which strived harsh weather has complicated patterns like fraise, which in fact is a result of its harsh life, but amazingly beautiful. Trying to draw out this emotional quality, this 18-cornered table is dyed in dark navy.

P33
HANA - Gachi Series (Round table, Armchair)
1985
ϕ1500×H700mm
W640×D540×H854 SH430mm

P34, 35
BRONX 1010 (Folding chair)
1992
Cassina-Interdecor Japan Inc.
W535×D480×H685 SH440mm

As brother series of BLITZ, it was commercialized in 1992. The folding structure was simplified.

P36, 37, 38, 39
BLITZ (Folding chair) TUNE
1981-94
Skipper s.p.a.
1995-
Cassina-Interdecor Japan Inc.
W530×D480×H710 SH435mm

P40, 41
CCC (Carbon Composite Chair)
1985-90
Yamaha Corp.
1994-
Okamura Corp.
W530×D510×H740 SH430mm

In a chair the construction and principle often appear on its expression and material. In the realization of figurative image, it often happens that material and technology decide the image, while the image enhances technical exploitations. Carbon (CFRP) used in this chair is still an expensive material and difficult to use in large quantities. In order to bring out its attributes and at the same time achieve market affordability, I have used this material in combination with beech wood, simulating structural durability on CAD. Face materials and finishings vary.

P42, 43
ASCEND (Stacking chair)
1987
Houtoku Co., Ltd. Artifort
W475×D510×H770 SH420mm

The success of this type of chair depends heavily on market costs. Therefore taking this project as an opportunity, the automatic bender and the automatic welder were introduced and the production line renewed in the manufacturing factory.

P44
ACTON (Sofa)
2000
Iyobe Kogeisha Co., Ltd. Okamura Corp.
W770-1320×D767×H750 SH400mm

The tablet can turn and you can work with mobile devices. One type comes with casters on the back legs, and another comes with placage attached to the side of armrests and backrest.

P45
FLESSO (Sofa)
2000
Iyobe Kogeisha Co., Ltd.
W729-1289-1849×D700×H720 SH410mm

The tablet can turn and you can work with mobile devices. Chairs can be jointed, and casters on its back legs enables the chair to move and change the layout.

P46, 47
ECHO (Chair)
1991
Minerva Co., Ltd.
W460×D517×H782 SH435mm

Frames are narrowly structured in order to use various wood materials because only thin material can be obtained depending on the type of wood. Compact figure and comfort are sought for.

P47
CARBON CHAIR (Prototype)
1988
W470×D520×H785 SH435mm

Expanding the idea from the tennis racket design I was doing at that time, I tried to convert the technique of tennis racket production into this chair.
This chair is formed by internal pressure of carbon graphite sleeves and has gut on its back. Back then due to high costs for dies and other facilities, it ended with the stage of prototype.

P48
ARIA SERIES (Chair)
1992
Iyobe Kogeisha Co., Ltd.
Sedie Friuli s.r.l.
W550-575×D552-560×H900 SH425mm
W1000-2000×D1000×H700mm

A breath of contemporary air has been taken into the classical expression. A chair and a table with expansive scale. Glossy warmth of hard maple wood and its form with tense curves require the high-skilled craftsman's hands.
Going over between drafts and prototypes, it was finally manufactured using copping machines and NC routers.

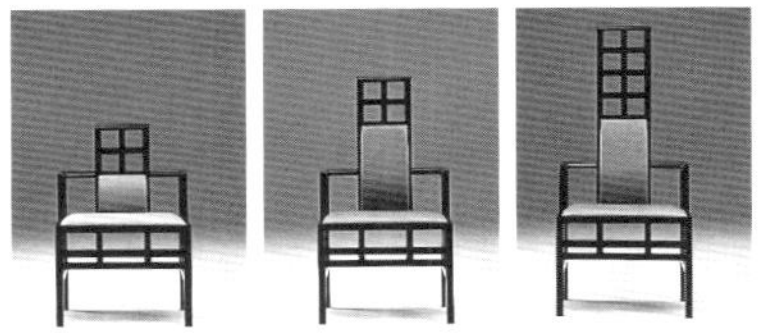

P49
JACKSONS (Chair)
1986
Takashimaya Kosakusho Co., Ltd.
W540×D480×H825-1030-1235 SH430mm

The subject here is grid. Cannot help approaching towards Mackintosh and Hoffmann, but I remember they were affected by Japonism and designed in that style. This is an interactive 'East meets West'.

P50, 51
SHIBA PARK TOWER / ENTRANCE HALL
2001
Mitsui Real Estate / Mitsui Mansion Project
Architectural desgn : KAJIMA DESIGN
Furniture manufacturing : Iyobe Kogeisha Co., Ltd.

This is a tower block in the city center. The 200 square-meter entrance hall has a 5 meter ceiling height, with natural color scheme and finishing material such as diatomite. At the north end there is a dark-colored Wenges flooring. The light-colored sofa upholstered in leather has high armrests and backrests. It delivers the airiness of space and at the same time it moderately hides the sitters, acting as a screen and creating a calm atmosphere.

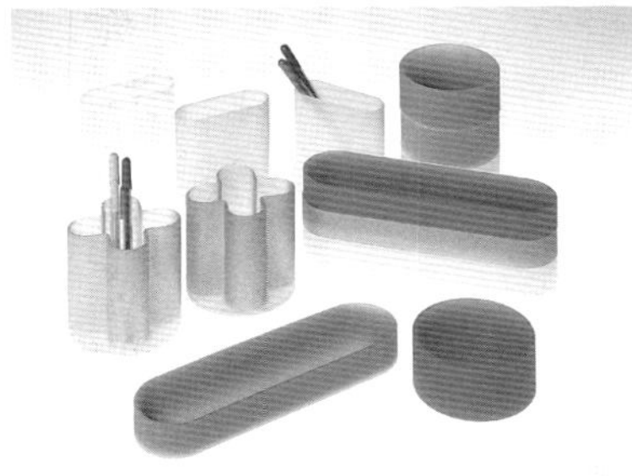

P52, 53
SILICOMBO (Deskware)
2000
Gallery Interform
Pen stand : W100×D100×H87mm, Pen tray : W240×D66×H42mm, Multipurpose tray : ϕ100×H42mm

An interesting material, silicone is soft, friendly and flexible. Soft touch stationary of solid form and lucent color like jellybeans. I hope to bring a breath of fresh air on to cold office desks.

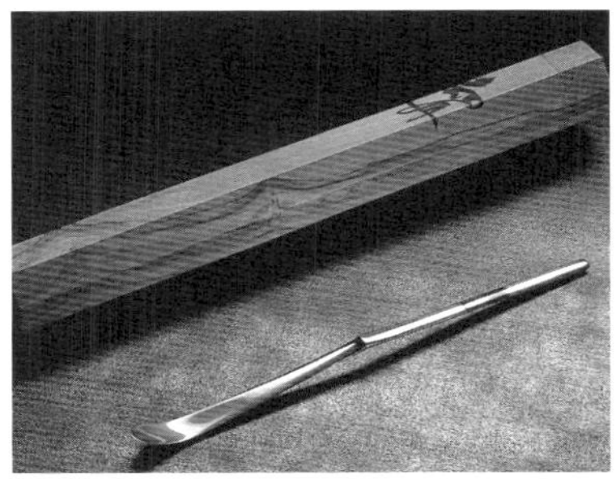

P54, 55
TITAN (Tea ceremony spoon)
1993
Takenaka Seisakusho Co., Ltd.
W10×L180×H10mm

An attempt to blend traditional crafts and contemporary design.
Comprehensive essence of Japanese culture is condensed in the tea ceremony.
Now that the material culture of mass production and mass consumption has settled down, it is exciting to reconsider the mind of tea ceremony through designing tea tools.

P56, 57, 58, 59, 60, 61
PLASMA TELEVISION　PX-42, PX-50
1998
NEC Corp.
Monitor : W1048×D91×H648mm
W1240×D107×H766mm

This media core can correspond to various image sources. Thin but large screen, high brightness and high quality pictures leading the next generation. The design is hard and sharp with aluminum. It doesn't assert its existence overmuch but suggests a new lifestyle as a multi purpose window for information. The development themes for this project were;
*How should its appearance be as an interface applied to various media.
*The monitor should not be too strong in its environment, however it should ensure high quality and high technology.
*It should suggest a new lifestyle.
*Creating a new market.
Of the two ideas presented, one, a movable wall-like system was too advanced to support the technical aspects.
The more orthodox idea was adopted, with separate displays, tuners and speakers. Not talkative but simple and hard form. The body is sharp and precise with extruded aluminum material.

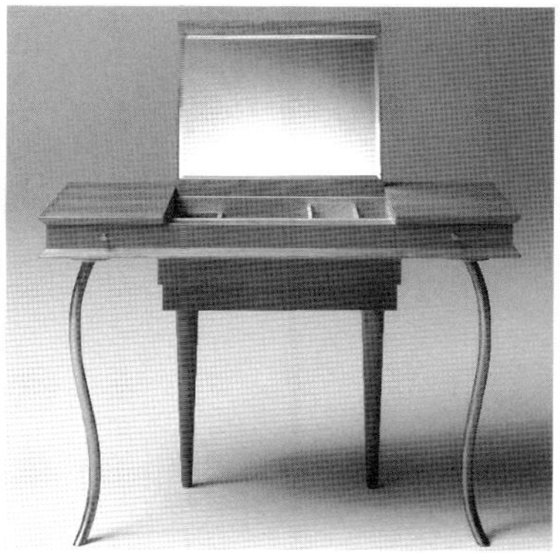

P62, 63
NEO MIND GEAR-DRESSER
1995
Shizuoka Furniture Cooperative Association
W1000×D485×H720-1120mm
ϕ490×H1400mm

The lifestyle of women has changed with generations, women today are craving for a way of free living. With the theme of "Mirror furniture suitable for contemporary women" I developed 7 types of furniture working with dresser makers of the Shizuoka Furniture Cooperative Association with design directive purposes in mind.

P64
MUG CUP
1992
MTG Design
ϕ75×H90mm

The mug cup with bar codes as a pattern.

P65
BLENDY　(Instant coffee)
1986
Ajinomoto General Foods Inc.
Graphic design : Shin Matsunaga
ϕ99×H178, ϕ94×H127, ϕ69×H94mm

They look nice when stacked in a shelf for display at a store, but their existence on a table and the messages delivered are more to note. Then after finishing the bottles, these bottles with wide-open mouths still have another use.

P65
MARIM　(Powdered cream)
1987
Ajinomoto General Foods Inc.
Graphic design : Shin Matsunaga
ϕ96×H137mm, ϕ96×H96mm, ϕ77×H79mm

P66
MALAY　(Trays, Bowls, Salad servers)
1988
Central Market Malaysia　Central Market Japan Co., Ltd.
ϕ120-300-350-400-600mm

Adding skilful job to warm wood material, I inspired a new life to it and expressed the freshness in our life. I assembled the mosaic of grain cut teakwood and gumwood from Malaysia and made the lathed trays, bowls and salad servers.

P67
GLASS TRAY
1975
Tachikichi Co.,　Adam&Eve
W215×D215×H35-45mm

Pressed glass trays in black and clear.
Two types of tray nest and work as lids.

P67
NEWS (Whisky bottle)
1983
Kirin-Seagram Ltd.
Graphic design : Shin Matsunaga
ϕ82×H130mm, ϕ87×H175mm, ϕ87×H270mm

It reflects the air of the generation, and delivers clear messages that show the inside.

P68
KAGA STICK GREEN TEA
1991
Maruhachi Seichajo Ltd.
Graphic design : Yukimasa Okumura
ϕ84×H116mm

This package carries the message of the contents and at the same time it is a re-usable container for another pack of tea. It is a package which conveys the sender's heart.

P69
Japanese Green Tea House ISSYO, Kanazawa

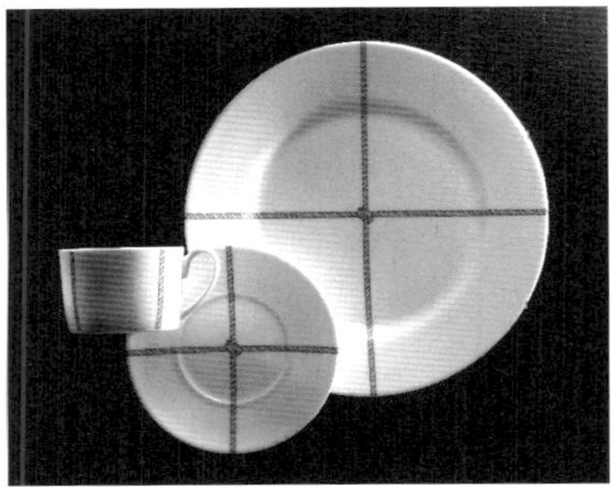

P70
TOKYO XII (Cup, Saucer, Plate)
1991
MTG Design

Cup : ϕ88×H60mm, Saucer : ϕ157×H15mm
Plate : ϕ270×H26mm

The era of design as business has ended and it has become the time of design as culture in a global sense.
The needs for the new generation is the joy for one's mind which transcends function and shape. Matching the vibe of sensitivity to the beat of new life, the keyword 〈Tokyo Twelve〉 was planned.

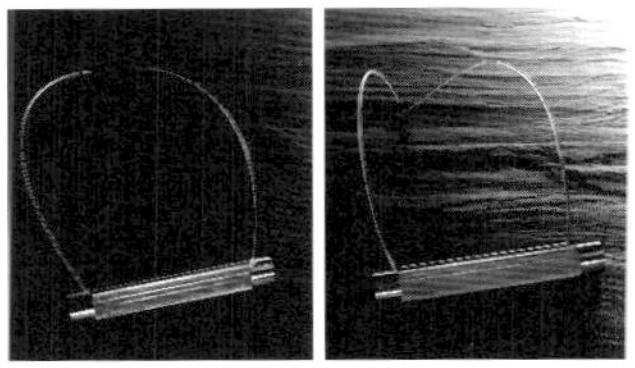

P71
METALMESH NECKLACE
1983
Limited Edition
ϕ20×L130mm

Metal mesh made of stainless steel and prosphor bronze is genuinely beautiful. I made a neckpiece with this material and exhibited it in an accessory show.

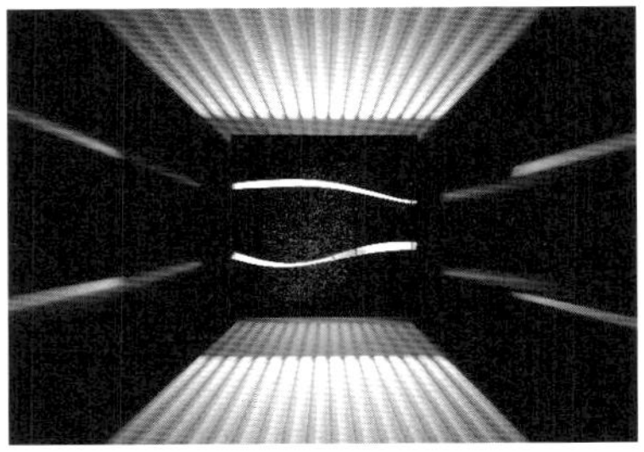

P72
GEN SERIES (Bracket)
1988
Yamagiwa Corp.
W450×D100×H120mm
W240×D100×H240mm

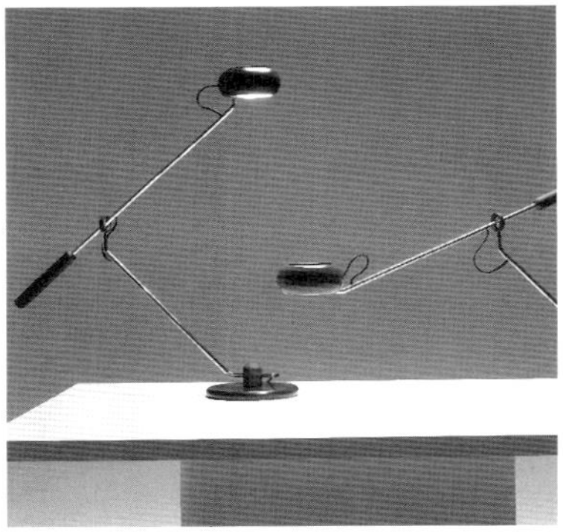

P73
RUACUL (Personal light)
1978
Yamagiwa Corp.
H440-760mm

As a personal light, the source of light can be set on the ceiling, on the wall or on the floor. In order to face any direction flexibly, each joint has a movable mechanism, and it is also intended to display its interesting changes.

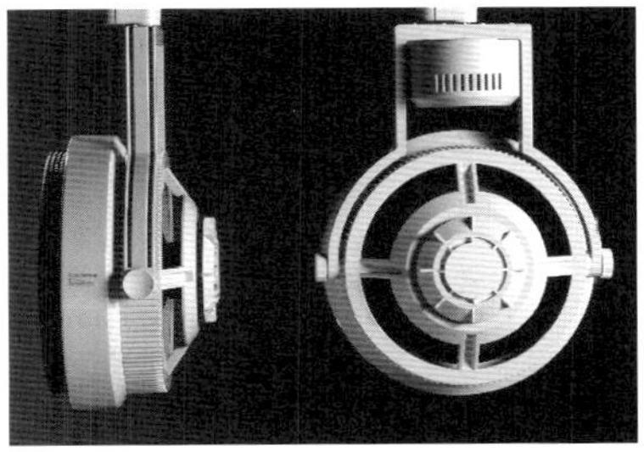

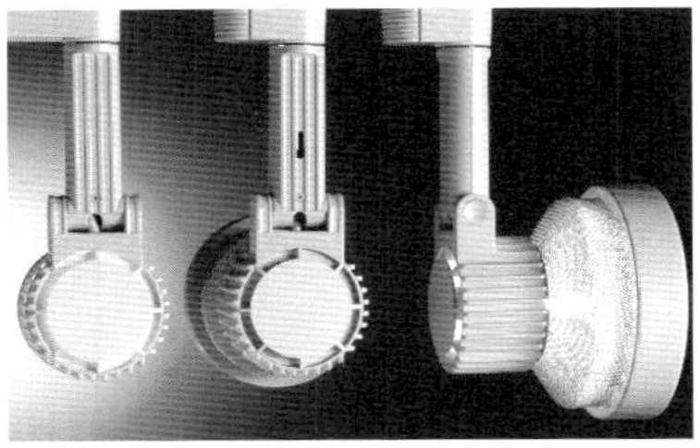

P74
ML SPOT LIGHT
SUPER, HYPER
1988
Yamagiwa Corp.
ϕ80-105-140-175-286mm

In order to achieve various luminous diversity in the lighting for commercial and retail facilities, this fixture uses halogen (100V), low-voltage halogen, diclo, HQI, HID, Metahara and other various sources, systematized to share the same elements in order to form a common rhetoric.

P74
RUACUL (Personal Light)
1978
Yamagiwa Corp.
H590-1500mm

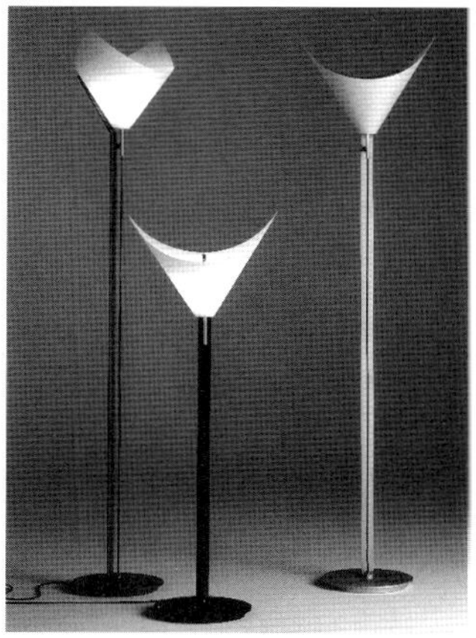

P75
ARIAKE (Floor stand lamp)
1983
Bushy Co., Ltd.
ϕ300×H1280-1880mm

I like the softness and the transitive light through Japanese paper. With the image of "Ariake", the dawn with the moon still in the sky, it implies abstracted Japanese images without depending solely on traditional materials.

P76
MODERN-JU (Nest of boxes)
1981
Bushy Co., Ltd.
W150-200-250-300×D150-200-250-300×H212mm

Once a year we cook special delicacies for the new year. To spend the new year's day at our aged parents' house, we make this meal and serve it in a set of Jubako, nest of boxes. I had been using the one that I had had for a long time, but I wanted a more simple one. For important occasions, happy moments and times for celebrating something, I like to take this out and use it more often.

P77
KASANE (Nest of tables)
1982
Bushy Co., Ltd.
W460-530-600×D400×H310-345-380mm

P78
KINU SERIES (Cabinet)
1982
Arflex s.p.a.
W85-125-165-205-2450×D500×H∞mm

The conversion from hand work to mechanical work. By the manufacturer's request, I used "CRESCENDO" of urushi lacquer finished cabinet as a model, and by changing from hands to machines, lacquer to polyester and polyurethane finish and analog methods to systematic methods, I re-designed the cabinet and added the table and the chairs to the series. I chose dyed silk tones of whitish brown, pink grey, rusty red, purple and black, and tried different combinations of shape, resulting in a somewhat sweet taste owing perhaps to the difference in the finishings and details.

P79
KINU SERIES (Chair)
1982
Arflex s.p.a.
Table : W1800×D1000-1300×H700
ϕ1300×H700mm
Chair : W540×D560×H870 SH460mm

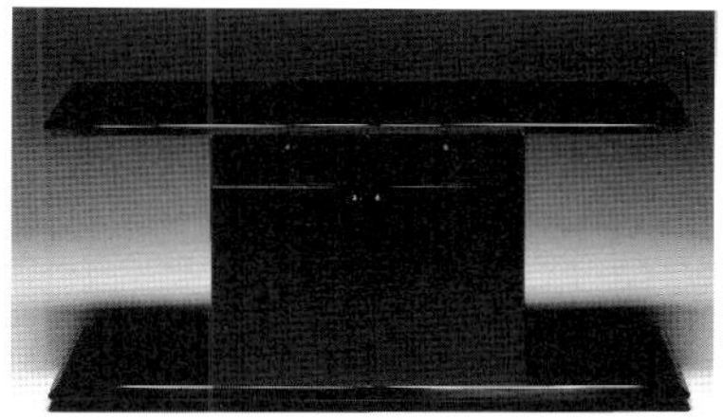

P80, 81
CRESCENDO A, B (Cabinet)
1981
Bushy Co., Ltd.
A,B : W1606×D496×H740mm
C : W826×D496×H1260

The best time of Japanese lacquer has passed. With the relationship between the image, the material and the shape that has lived through the ages in mind, this is one piece of furniture which I have been designing to revive the lacquer and to present it in contemporary furniture. I chose urushi tallow finish and dry lacquer finish out of numerous lacquering methods and added the stoic but tranquil taste and glossy coating to its form. Purple is said to be one of unstable colors in lacquer but this one shouldn't be excluded. It was lucky to have met a skilful lacquerer. This is the world of handiwork where a skilful craftsman is indispensable for a large piece.

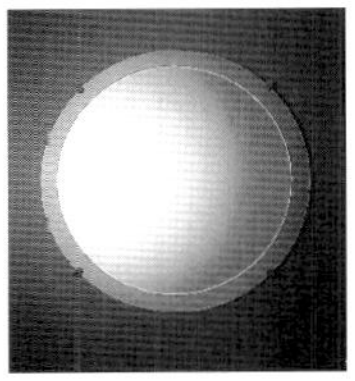

P82
SANITARY ACCESSORY MIRROR - MK Series
1996
TOTO Co., Ltd.
W350×H1300mm, W400×H500mm, W500×H650mm
ϕ450-650-800mm, W800×H600mm
W300×H450mm, W400×H600mm

The changes in contemporary lifestyle have been affecting the lavatory space. What is needed now is not only basic functions such as washing and grooming. It has become a multi-purpose space with added value depending on the user's interests, for example bringing in exercise machines. The mirror has become not only a tool but an equipment and further a part of the interior. The mirror shouldn't be too flashy and should be a part of the space.
Not to mention its function is important but it should be flexible to match any space. This mirror itself, based on contemporary life, I designed to create "a fixture of daily standard necessities" that needs meet the cost.

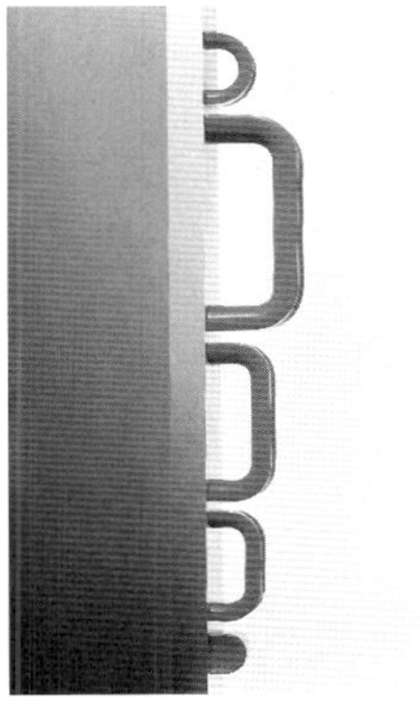

P83
PUSH & PULL
1974
Arflex Japan Ltd.
W24×D16×H21mm, W42×D13×H26mm, W66×D16×H29mm, W93×D20×H35mm, W126×D24×H52mm

Topologically changed, there are 5 sizes and 6 types from a small drawer handle to a doorknob. Comes in orange, yellow, white, green and black.

P84, 85, 86, 87
TSURUMI TSUBASA BRIDGE
1994
Metropolitan Expressway Public Corporation
Place : Metropolitan Expressway Bay Shore Route-From Ougishima to Daikoku Wharf, Tsurumi-ku, Yokohama.
Owner : Metropolitan Expressway Public Corporation
Structural engineering : Shinnihongiken, Yachiyo Engineering Co., Ltd.
W38×L1020×H180m

I was involved in this project for 6 years from the viewpoint of a landscape designer. We as designers should create and save beautiful landscape and good living environment. And now we have the chance to be involved in projects for public buildings, bridges, roads and parks, riverside maintenances and the like.
Awarded Tanaka Prize at Japan Society of Civil Engineers, and the Yokohama Landscape Award.

P88
NEX SERIES
1999
Uchida Yoko Co., Ltd.

This demountable office desk system is made of recycled aluminum, designed with environmental friendliness and LCA in mind. Coordinated with the chair and the panel, it is possible to produce office spaces as a whole and to accommodate various work styles. It fulfills landscape beauty as an object, and at the same time as a group.

P89, 90
NEX SERIES PARTITION
1999
Uchida Yoko Co., Ltd.

This partition panel is made of aluminum material which can be recycled.
Coordinated with the desk and the chair, they work as one system and can produce office spaces as a whole. These interactive lucent sliding doors and flexible panels can produce various spaces. It fulfills landscape beauty as an object, and at the same time as a group.

P91
LOBBY CHAIR FOR OZONE
1994
Okamura Corp.
W1855×D640×H400-760　SH400mm

The lobbies situated in the entrance and on each floor of Tokyo Gas Ozone Building.
They compose a hard and clear atmosphere distinct from the one of surrounding showrooms for furniture and fabrics.

P92
NIKI CLUB NASU PART 1
1986
Architectural design : Akira Watanabe
Furniture manufacturing : Yamaha Corp.
Interior Center Co.

Talkative surface coverings and furniture do not fit in this space ruled by the existence of stone, constructed with warm Otani stone and thick pine wood with chipped finish. The trace of hand has been left on the surface of stone. To correspond with the quality of the rich space of this lounge, a large heavy solid table and high backrest chairs are set on the axis. When you listen to the jabbling of the approach and the noise from the river, an open space is created where you can see great mountains through the trees.
Facing this vista are pairs of lounge chairs. Lead lines and eye lines are taken into consideration and wide armrests work as side tables. This entrance is spacious but looks compact. Contrary to the open clearness of the day, lights in the hands will glow sensitively and calmly when the sun has fallen.

P93
JAPAN NATIONAL OIL CORPORATION
TECHNOLOGY RESEARCH CENTER
1988
Architectural design : Nihon Sekkei Co., Ltd.
Furniture manufacturing : Iyobe Co., Ltd.
Tendo Co., Ltd.　Houtoku Co., Ltd.
Daichi Co., Ltd.

It is a facility which has a study room as well as hotel and restaurant function.

P94
RACK A, B (For CD)
2001
S.B.Decola & Building Materials Co., Ltd.
A : W650×D175(314)×H1230mm
B : W810×D175(314)×H1070mm

This is a CD rack made of lucent materia and metallic material.

P95
SYSTEM KITCHEN SUPER LEGACESS-L
2001
TOTO Co., Ltd.

The kitchen that can match Japanese culture which enjoys Japanese, Western and Chinese food. It is handy, easy to clean and good to store things. Developed under the themes, how to cope with its user's needs, and how to create a rich dinning space.

P96, 97
SYSTEM BATHROOM - Furopia KV Series
2001
TOTO Co., Ltd.
W2100×D1650×H2310mm
W1650×D1650×H2310mm

Not only in housing projects but also in single houses, the construction process of bathrooms has been changing from the conventional one to the prefabricated process, and the variation in sizes and qualities have been increasing. They have been highly systematized and the quality of living spaces has been reconsidered. Among these this type is quite spacious inside, and with the concept of "depth and brightness", it presents a high quality material texture and has a lot of ideas for the storage of bathing goods, and consideration for universal design.

P98
ZAZA - Motomi Kawakami Collection
1993
TOTO Co., Ltd.
Automatic sensored washbasin and flushing bulb lavatory

It has a texture like it was carved out of a block of granite and ground on the surface, with present technology inside and sharp and simple form outside. I used the same texture and material for the floor and walls in the public space, and tried to accomplish the integration of hard but comfy furniture and its environment.

P99
ZAZA - Motomi Kawakami Collection
1993
TOTO Co., Ltd.
Sensored flushing bulb stool lavatory
Washbasin with automatic mixing faucet and pedestal

P100
WASHBASIN PROJECT (Scale model)
1994
W800×D450×H1900mm

The water filled in the basin glows glamourously lit from the bottom. In its cold shape, it delivers "profundity".

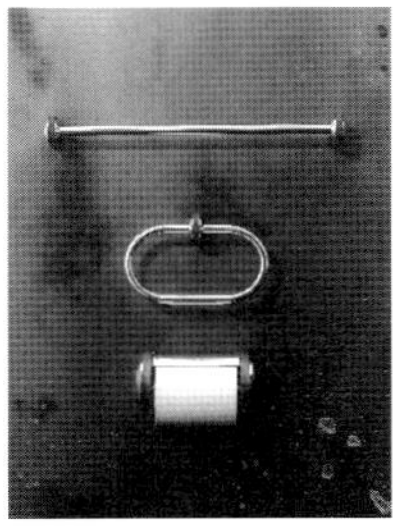

P100
SANITARY ACCESSORY - Material Design Series
Glass, Aluminum, Wood
1999
TOTO Co., Ltd.

(Aluminum) Fingertip toilet roll holder : W164×D120×H90mm, Towel bar : W444×D68×H38mm, Towel ring : W194×D68×H125mm
(Wood) Fingertip toilet roll holder : W170×D122×H99mm, Towel bar : W 484×D75×H57mm, Towel ring : W161×D75×H184mm
(Glass) Fingertip toilet roll holder : W172×D122×H100mm, Towel bar : W444×D74×H48mm, Towel ring : W214×D74×H141

P101
SANITARY ACCESSORY - MK Series
1996
TOTO Co., Ltd.
W400×D115-134×H115-134mm

Double toilet roll holders come with the shelf made of aluminum, glass, porcelain, natural wood, and marblite and so on. Universal design functions such as easy changing of toilet rolls and cutting with one hand are taken into consideration.

P102,103
PEDESTRIAN FURNITURE (Project)
1995

It is basically constructed on a 60cm grid. Two dimentional parts such as floor units, gutters, lawn units, and the like are prepared, and also three dimentional parts such as bollards, water fountains, rubbish bins, planters, benches, tables, sign poles, lamp poles, shelters, pergolas and others. Registered design.

P104, 105
PERGOLAS & SHELTERS - KD Series
1994
Hokusho Co., Ltd.
Basic Span : 2.5m, 3.0m, 3.5m

This KD series is a system composed of pergola, shelter and street furniture with the concept of the micro view, that sticks to the texture and details of materials, and the macro view, that concerns the design as elements in the space. The "columns", "beams" and "bars and roofs" are the main components for the pergola and shelter. This is the system which enables to produce appropriate spaces by assembling them variously to match with the image from architectural plans. One of the main components "columns" are assembled as a 4 part architecture in a socket style, and choosing the design and the color scheme freely out of 10 types of shape and 4 colors, it can respond to architectural intentions widely. There are also wood-affixed type and trellis type that plants can be bound to according to the locations. "Beams" are extruded aluminum parts whose outer dimension is 120mm with rail gutters in 4 directions and added flat joints, they fix pergola bars and main building of shelter and other components together.
These beams jointed with cast aluminum joints can be connected vertically, horizontally and diagonally with columns and have further freedom of layout.
At the same time, it can be expanded to street furniture (benches, signs, fences, etc.) with the same materials. Californian redwood is used for the wood part. 6 types of parts can be selected for "bars and roofs" according to beam frame shape. The roofboard and rafter have been incorporated into a single panel structure, by using big panels of OSB wood in order to reduce cost and constructing time at the site, and to stabilize the quality and the supply with stocks.

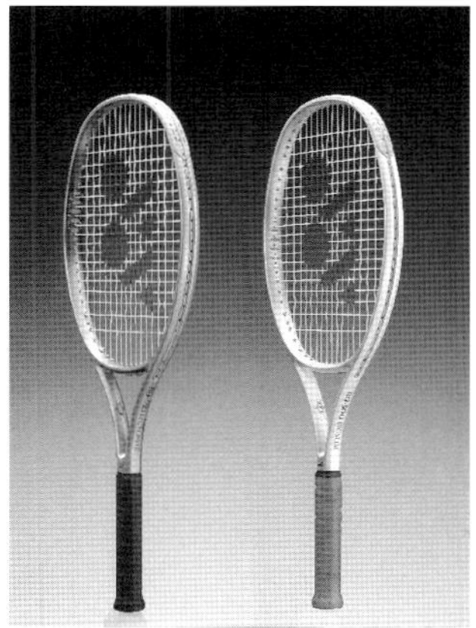

P106
TENNIS RACKET - RQ Series
1988
Yonex Co., Ltd.
W225×D30×L680mm

With the concept of sometrics, the proportion of the sweet spot against the racket size is made bigger. This shape will become an identity of the manufacturer. I designed it at the time of changes of materials from plywood and hard aluminum to plastic. At that time, new materials such as graphite, kevler, boron and silicone were tried out.

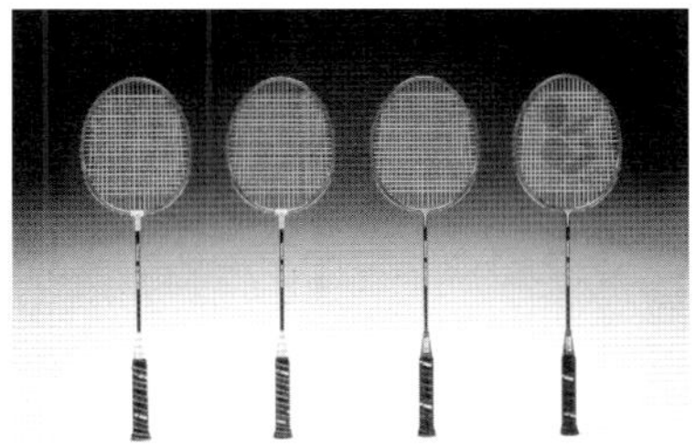

P107
BADMINTON RACKET - Carbonex-8
1986
Yonex Co., Ltd.
W205×D10×L570mm

This is the highly bouncy racket made of carbon shaft and aluminum alloy.
Badminton rackets later move on to all carbon.

P108
INTERFLEX (Desk clock)
2000
Gallery Interform
W80×D80×H80mm

The design expresses the charm of round gravels spread on the riverside, and an appropriate sense of weight.

P109
INTERFLEX (Wrist watch, Eco drive)
2000
Gallery Interform
W34×D40×H7.4mm

This is a wrist watch which has beautiful plasticity in its thin body. It changes solar energy into electricity and ticks time. When fully charged it can keep on moving in the dark for about 6 months, and even a 20 lux light like interior fluorescent lamp is enough to charge the second battery. Harmful substances are not used for the battery and this watch is quite ecological.

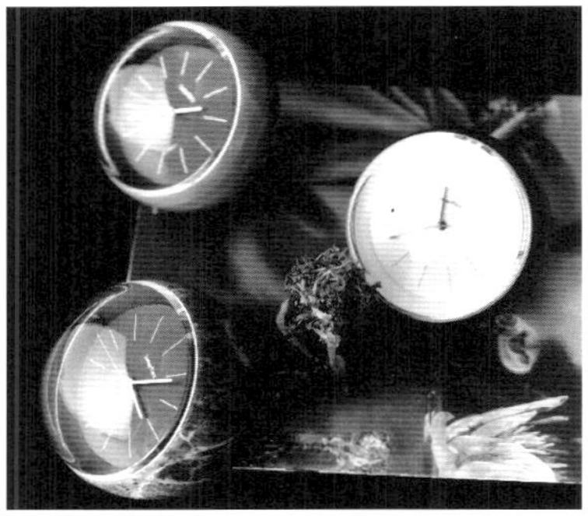

P110
INTERFLEX (Desk clock, Prototype-granite)
2001
Gallery Interform
W80×D80×H80mm

Using granite and marble for its body, it is just a clock of rock.

P111
INTERFLEX (Wall clock)
2000
Gallery Interform
ϕ204×D32mm

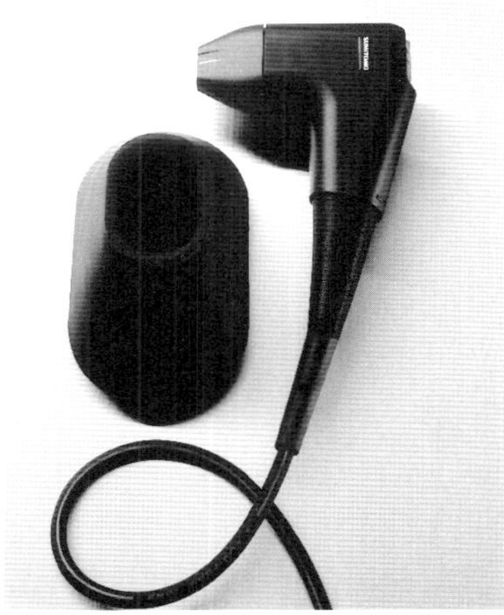

P112
HAND SCANNER
1976
Sumitomo Electric Co., Ltd.
W189×D23×H88

OCR [Optical Character Recognition] of hand-held type. Interface technology has been progressing and the size has become compact.

PHOTOGRAPHS

Akira Tabuchi
81

Atsusi Sibata
56, 57

BALLO & BALLO
7, 78, 79

Bitetto e Chimenti
38

Hajime Inoue
95, 97

Hiroyuki Hirai
94

Hoichi Nishiyama
84, 86

Joji Kanda
109

Katsuhisa Kida
35

Koichi Inakoshi
5, 8, 9, 10, 22, 23, 24, 25, 26, 27, 29, 52, 53, 54, 64, 68, 89, 96, 108, 110

Masahiro Yoshio
69

Masami Harada
62, 63

Mitsumasa Fujitsuka
50, 51, 92, 98, 99

Nacasa & Partners
34, 36, 66, 70

Osamu Murai
93

Osamu Nagahama
13

Pietro Carrieri
20

Syozo Nakamura
28

Teiichi Kayama
58, 59, 60

Tetsuhiko Kudou
100

Yoshio Shiratori
12, 14, 15, 16 , 17, 18, 19, 21, 30, 31,32,33, 37, 41, 42, 43, 46, 47, 55, 65, 67, 71, 73, 74, 75,77, 80, 82, 83, 87, 91, 111, 112

Yoshiteru Baba
100, 101,102, 105

Cassina-Interdecor Japan Inc.
39

Soen Co., Ltd.
44, 45, 48

Takashimaya Kosakusho Co., Ltd.
49

Yamagiwa Co., Ltd.
72, 74

Bushy Co., Ltd.
76

Metroporitan Expressway Public Corp.
85

Uchida Yoko Co., Ltd.
88, 89, 90

Yonex Co., Ltd.
106, 107

CASA VOGE
38

Arflex Japan Ltd.
13

Kawakami Design Room
47, 61, 103, 104

MOTOMI KAWAKAMI

KAWAKAMI DESIGN ROOM
6-14 Ohyama-cho Shibuya-ku
Tokyo 151-0065 Japan
TEL. ++81-3-3481-9891 FAX. ++81-3-3481-9892
E-MAIL. gachi@ba2.so-net.ne.jp